P9-CKY-263

ISTITUTO NAZIONALE DI ALTA MATEMATICA
FRANCESCO SEVERI

SYMPOSIA MATHEMATICA

VOLUME XXX

ACADEMIC PRESS LONDON AND NEW YORK 1989

Published by Istituto Nazionale di Alta Matematica Francesco Severi Roma

Pubblicato con un parziale finanziamento
del Consiglio Nazionale delle Ricerche

©1989 by Istituto Nazionale di Alta Matematica Francesco Severi

Distributed throughout the world by

Academic Press Inc. (London) Ltd

24-28 Oval Road, London NW1 7DX

ISBN 0.12.612230.X

Printed in Italy
by Tecnoprint – Via del Legatore 3, Bologna – 1989

CONTRIBUTORS

A. ACKER Department of Mathematics and Statistics, Wichita State University, Wichita, Kansas 67208, Usa.

F. J. ALMGREN JR. Department of Mathematics, Princeton University, Princeton, New Jersey 08544, USA.

C. BANDLE Mathematisches Institut, Universität Basel, Rheinsprung 21, CH-4051 Basel, Switzerland.

E. FABES School of Mathematics, University of Minnesota, 206 Church St. SE, Minneapolis, Minnesota 55455, USA.

R. FINN Mathematics Department, Stanford University, Stanford, California 94305, USA.

N. GAROFALO Dipartimento di Matematica dell'Università, piazza di Porta S. Donato 5, 40127 Bologna, Italy.

B. KAWOHL Institut für angewandte Mathematik, Universität Heidelberg, Im Neuenheimer Feld 294, D-6900 Heilderberg, BRD.

G. HEADY Department of Mathematics, University of Western Australia, Nedlands, WA 6009, Australia.

E. LIEB Department of Physics, Jadwin Hall, P. O. Box 708, Princeton New Jersey 08544, USA.

S. MARINE-MALAVE School of Mathematics, University of Minnesota, 206 Church St. SE, Minneapolis, Minnesota, 55455, USA.

L. MODICA Dipartimento di Matematica dell'Università, via Buonarroti 2, 56100 Pisa, Italy.

L. E. PAYNE Department of Mathematics, Cornell University, Ithaca, New York 14853, USA.

G. PHILIPPIN Département de Mathématique, Université Laval, Cité Universitaire, Québec, Canada G1K-74P.

S. SALSA Dipartimento di Matematica del Politecnico, via Bonardi 9, 20133 Milano, Italy.

J. SERRIN School of Mathematics, University of Minnesota, 206 Church St. SE, Minneapolis, Minnesota 55455, USA.

I. STAKGOLD Department of Mathematical Sciences, University of Delaware, 501 Ewing Hall, Newark, Delaware 19716, USA.

T. SUZUKI Department of Mathematics, Faculty of Science, Tokyo Metropolitan University, 2-1-1, Fukazawa, Setagaya, Tokyo, Japan.

FOREWORD

A conference with the title «Geometry of solutions to partial differential equations» was held in Cortona, Italy, from June 16 to June 21, 1988, under the auspices of the Istituto Nazionale di Alta Matematica. The conference took place in the Palazzone, a XVI-century castle now owned by the Scuola Normale Superiore, and was attended by some fifty participants. The organizing committee was made up of Luis Caffarelli (I.A.S. Princeton), Bernhard Kawohl (Universität Heidelberg), Luciano Modica (Università di Pisa) and myself.

The conference was devoted to new developments in PDE, especially those based upon geometric properties of solutions. Topics such as symmetries, geometry of level sets, rearrangements – which are now receiving the attention of many – were dominant themes. The conference consisted of invited one-hour plenary addresses and a few contributed papers. This volume collects reports of the plenary addresses.

One of the speakers was Hans Lewy. For a long time Hans Lewy was involved in the problem of counting the umbilical points of a closed two-dimensional surface – a problem formerly settled by Carathéodory in the analytic case. He felt he had a solution in hand in the much more difficult, essentially different case of finitely differentiable surfaces and spoke in Cortona about this matter – thus generously spreading his ideas, though regretting, as he modestly claimed, to be unable to present a detailed proof. He also participated in many informal discussions and social events, always showing his exceptionally deep insight and wide experience as well as his admirable enthusiasm. The stimulating and warm atmosphere, that people enjoyed in Cortona, largely resulted from him. Hans Lewy died on August 23, 1988. I am sure that all of us, who had the privilege of listening to Lewy's last talk, would like to have this volume dedicated to his memory, with the hope that it may contribute to the celebration of one of the most distinguished mathematicians of our age.

Firenze, 1 January 1989 G. TALENTI

CONTENTS

GEOMETRY OF SOLUTIONS TO PARTIAL DIFFERENTIAL EQUATIONS

Proceedings of a conference held in Cortona, Italy, 16-21 June 1988

Editor: Giorgio Talenti, Università, Firenze

dedicated to the memory of Hans Lewy

THE BERNOULLI FREE-BOUNDARY PROBLEM-UNIQUENESS AND ELLIPTIC ORDERING OF SOLUTIONS

ANDREW ACKER

1. INTRODUCTION

The purpose of this paper is to present a reasonably general method (or framework) for studying the uniqeness question for solutions of some free-boundary problems in elliptic PDEs. This method also provides convergence theorems for (a particular version of) the well-known trial free-boundary method for the successive approximation of solutions to free-boundary problems. In both cases, the results follow from the study of parametrized families of solutions which depend continuously and monotonically on the parameter. For expliciteness, our discussion will be in the context of the interior Bernoulli free-boundary problem to be stated below. Our explicite uniqueness and convergence results will be for a relatively arbitrary geometric situation in two dimensions and for the convex situation in $n \geq 3$ dimensions.

1.1. INTERIOR BERNOULLI FREE-BOUNDARY PROBLEM. (See Fig. 1).

Let G denote a bounded, simply-connected C^2-domain in $\mathbf{R}^n (n \geq 2)$ with boundary $\Gamma^* = \partial G$. Given $\lambda > 0$, we seek a simply connected subdomain D_λ of G (or its boundary $\Gamma_\lambda = \partial D_\lambda \subset G$) such that

$$(1.1) \qquad |\nabla U_\lambda(p)| = \lambda \text{ on } \Gamma_\lambda,$$

where U_λ denotes the capacity potential in the annular domain $\Omega_\lambda := G \backslash \bar{D}_\lambda$, i.e. U_λ solves the boundary value problem

$$(1.2) \qquad \Delta U_\lambda = 0 \text{ in } \Omega_\lambda, U_\lambda(\Gamma_\lambda) = 0, U_\lambda(\Gamma^*) = 1.$$

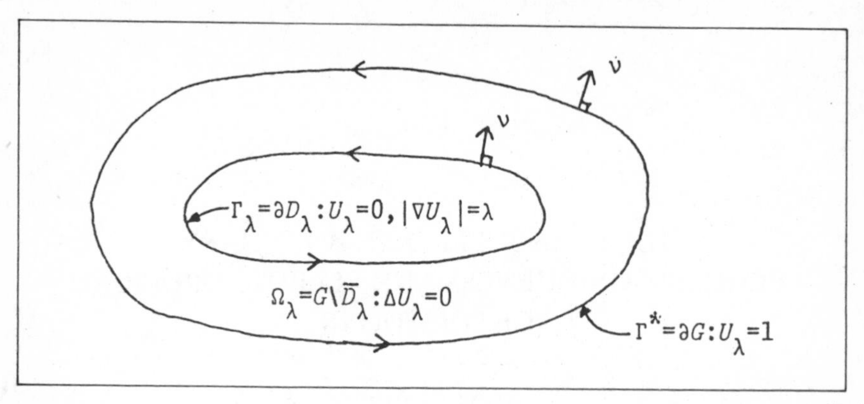

Fig. 1.

1.2. REMARKS. (a) In the context of ideal fluids $(n = 2), U_\lambda$ is the stream function of a flow in Ω_λ; and Γ_λ is the flow-boundary of constant pressure.

(b) For $n = 2$, existence results are due to Lavrentiev [La], Beurling [B], Daniljuk [D]. Beurling's result is stated in §3. Existence and regularity results in the n-dimensional case are due to Alt and Caffarelli [A-C].

(c) To the author's knowledge, there are no uniqueness results in the literature for Problem 1.1, even in 2 dimensions. However, the solution of the corresponding exterior Bernoulli free-boundary problem is unique in the starlike case (see [T] and [A1]).

1.3. GENERAL NOTATION. Let $\underline{X}$ be the set of all $(n-1)$-dimensional hypersurfaces Γ of the form $\Gamma = \partial D$, where D is a simply-connected subdomain of G. Given $\Gamma \in \underline{X}, U(p)$ denotes the capacity potential in the annular domain $\Omega := G\setminus\overline{D}$. The notation $D_\lambda, U_\pm, \tilde{\Omega}$, etc. is defined analogously for surfaces $\Gamma_\lambda, \Gamma_\pm, \tilde{\Gamma} \in \underline{X}$. For $\Gamma_1, \Gamma_2 \in \underline{X}$, we write $\Gamma_1 \leq \Gamma_2$ (resp. $\Gamma_1 < \Gamma_2$) if $D_1 \subset D_2 (\bar{D}_1 \subset D_2)$. For any sets A, B, we define dist $(A, B) = \inf\{|p-q| : p \in A, q \in B\}$. For $\Gamma_1, \Gamma_2 \in \underline{X}$, we use $\Delta(\Gamma_1, \Gamma_2)$ to denote the smallest value $\epsilon \geq 0$ such that $D_1 \subset N_\epsilon(D_2)$ and $D_2 \subset N_\epsilon(D_1)$, where $N_\epsilon(\cdot)$ denotes the Euclidean ϵ-neighborhood.

2. ELLIPTIC ORDERING AND UNIQUENESS

Let I denote a positive open interval. For each $\lambda \in I$, let the $(n-1)$-dimensional hypersurface $\Gamma_\lambda = \partial D_\lambda$ be a solution of Problem 1.1 at the parameter value λ. Then the solution family $\Gamma_\lambda, \lambda \in I$ is called *elliptically ordered* (see Beurling [B]) if

$$\Gamma_\alpha < \Gamma_\beta \text{ whenever } \alpha < \beta \text{ in } I. \tag{2.1}$$

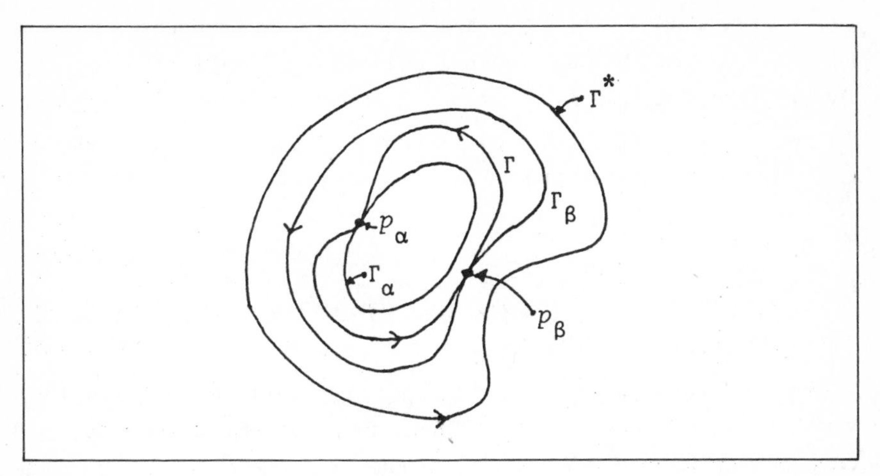

Fig. 2.

Thus, elliptic ordering means that the solution surfaces $\Gamma_\lambda, \lambda \in I,$ move outward toward Γ^* as λ increases. Our interest in elliptic ordering derives mainly from the following fact:

2.1. THEOREM. Given an elliptically-ordered family of solutions $\Gamma_\lambda, \lambda \in I,$ which are continuously varying in the parameter λ, let the annular domain R be the set-theoretic union of all the surfaces $\Gamma_\lambda, \lambda \in I$. For any $\lambda_0 \in I$, let $\Gamma = \partial D$ denote an arbitrary solution (at λ_0) such that $\Gamma \subset R$ and such that D contains the interior complement of R. Then in fact $\Gamma = \Gamma_{\lambda_0}$.

PROOF. (See Fig. 2). Let $\alpha \in I$ be maximum and $\beta \in I$ be minimum subject to the requirement that $\Gamma_\alpha \leq \Gamma \leq \Gamma_\beta$, and let $p_\alpha \in \Gamma \cap \Gamma_\alpha$ and $p_\beta \in \Gamma \cap \Gamma_\beta$. We have $\alpha \leq \beta$ due to the elliptic ordering. By the maximum principle, we have

$$U(p) \leq U_\alpha(p) \text{ in } \Omega, U_\beta(p) \leq U(p) \text{ in } \Omega_\beta,$$

where U denotes the capacity potential in $\Omega := G\backslash\bar{D}$. Since

$$U_\alpha(p_\alpha) = U(p_\alpha) = 0 = U(p_\beta) = U_\beta(p_\beta),$$

we conclude that

$$\alpha = |\nabla U_\alpha(p_\alpha)| \geq |\nabla U(p_\alpha)| = \lambda_0 = |\nabla U(p_\beta)| \geq |\nabla U_\beta(p_\beta)| = \beta$$

Thus $\alpha = \lambda_0 = \beta$ and $\Gamma = \Gamma_{\lambda_0}$.

2.2. COROLLARY. Let $\Gamma_\lambda, \lambda \in I$, and $\tilde{\Gamma}_\lambda, \lambda \in J$, be two continuously-varying, elliptically ordered families of solutions of Problem 1.1, and let $R = \cup_{\lambda \in I}\Gamma_\lambda$. Assume for a value $\lambda_0 \in I \cap J$ that $\tilde{\Gamma}_{\lambda_0} \subset R$ and that the interior complement $\tilde{D}_{\lambda_0}$ of $\tilde{\Gamma}_{\lambda_0}$ contains the interior complement of R. Then $\Gamma_\lambda = \tilde{\Gamma}_\lambda$ for all $\lambda \in I \cap J$. Therefore, a larger continuously-varying, elliptically-ordered solution family $\Gamma_\lambda, \lambda \in I \cup J$ is define by setting $\Gamma_\lambda = \tilde{\Gamma}_\lambda$ for $\lambda \in J \backslash I$.

3. EXISTENCE OF ELLIPTICALLY-ORDERED FAMILIES OF SOLUTIONS IN 2-DIMENSIONS

The key condition guaranteeing the elliptic ordering of a given, continuously varying solution family $\Gamma_\lambda, \lambda \in I$, in 2-dimensions turns out to be that $K_\lambda(p) < \lambda$ on Γ_λ for each $\lambda \in I$, where $K_\lambda(p)$ denotes the curvature of Γ_λ at the point $p \in \Gamma_\lambda$.

3.1. THEOREM. Let $n = 2$ in Problem 1.1, and let Γ_{λ_0} be a solution (at λ_0) such that $K_{\lambda_0}(p) < \lambda_0$ on Γ_{λ_0}. Then there exists a maximal, elliptically-ordered, locally Lipschitz-continuously varying family of solutions $\Gamma_\lambda, \lambda \in I$, containing Γ_{λ_0} as a member, such that $K_\lambda(p) < \lambda$ on Γ_λ for all $\lambda \in I$.

The remainder of this section will be devoted to the proof of Theorem 3.1, in which the following existence theorem due to Beurling [B] plays an important role:

3.2. THEOREM. (Beurling [B]). Let $n = 2$ in Problem 1.1. Given $\lambda > 0$, let $\Gamma_\pm \in \underline{X}$ be C^2 curves such that $\Gamma_- < \Gamma_+, |\nabla U_-| < \lambda$ on Γ_- and $|\nabla U_+| > \lambda$ on Γ_+. Then there exists a solution Γ_λ of Problem 1.1 (at λ) such that $\Gamma_- < \Gamma_\lambda < \Gamma_+$.

3.3. REMARK. Any solution Γ_λ of Problem 1.1 (with $n = 2$) is an analytic curve such that U_λ can be harmonically continued to an annular domain G_λ containing $\Omega_\lambda \cup \Gamma_\lambda$ (see Lewy [Le]). In fact, given a solution Γ_{λ_0}, one can show that there exist values $\delta_0, \epsilon_0 > 0$ such that for all solutions Γ_λ satisfying $|\lambda - \lambda_0| < \delta_0$ and $\Delta(\Gamma_{\lambda_0}, \Gamma_\lambda) < \delta_0$, the corresponding capacity potential U_λ can be harmonically continued to the domain $\Omega_{\lambda_0} \cup N_{\epsilon_0}(\Gamma_{\lambda_0})$.

3.4. DEFINITIONS. For any solution Γ_λ of Problem 1.1 and for $\delta < 1$, we define

$$\Gamma_{\lambda,\delta} = \{p \in G_\lambda : U_\lambda(p) = \delta\}, \Omega_{\lambda,\delta} = \{p \in G_\lambda : U_\lambda(p) > \delta\},$$

$$U_{\lambda,\delta}(p) = (U_\lambda(p) - \delta)/(1 - \delta) \text{ in } \bar{\Omega}_{\lambda,\delta},$$

where G_λ is the maximal annular domain containing $\Omega_\lambda \cup \Gamma_\lambda$ in which U_λ is harmonic. Notice that if $|\delta|$ is sufficiently small $(|\delta| \leq \delta_0(\lambda))$, then $\Gamma_{\lambda,\delta} \in \overline{X}$ and $U_{\lambda,\delta}$ is the capacity potential in the corresponding annular domain $\Omega_{\lambda,\delta}$. For $p_0 \in \Gamma_\lambda$ and $|\delta| < \delta_0(\lambda)$, we define

$$p_\delta = p_0 + \alpha\nu(p_0) \in \Gamma_{\lambda,\delta}, \tag{3.1}$$

where $\nu(p_0)$ is the exterior normal vector to Γ_λ at $p_0 \in \Gamma_\lambda$ and where $|\alpha|$ is minimum subject to (3.1).

3.5. LEMMA. Given a solution Γ_λ, we have

$$|\nabla U_{\lambda,\delta}(p_\delta)| = \lambda + (\lambda - K_\lambda(p_0))\delta + O(\delta^2) \tag{3.2}$$

as $\delta \to 0$, uniformly for all $p_0 \in \Gamma_\lambda$. Therefore, if $\lambda - K_\lambda(p) \geq C > 0$ on Γ_λ and $|\delta|$ is sufficiently small $(|\delta| \leq \delta_1(\lambda, C) < \delta_0(\lambda))$, then

$$|\nabla U_{\lambda,\delta}(p)| \geq \lambda + C\delta/2 > \lambda \text{ on } \Gamma_{\lambda,\delta} \text{ when } \delta > 0, \tag{3.3}$$

$$|\nabla U_{\lambda,\delta}(p)| \leq \lambda + C\delta/2 < \lambda \text{ on } \Gamma_{\lambda,\delta} \text{ when } \delta < 0, \tag{3.4}$$

PROOF SKETCH. Fix $p_0 \in \Gamma_\lambda$ and let ν denote the exterior normal vector to Γ_λ at p_0. Since $U_\lambda(p_0) = 0$ and $|\nabla U_\lambda(p_0)| = \lambda$, the Taylor Theorem shows that $\delta = U_\lambda(p_\delta) \simeq \lambda(p_\delta - p_0) \cdot \nu$, and hence that

$$\begin{aligned} |\nabla U_\lambda(p_\delta)| &\simeq \lambda + (\partial^2 U_\lambda(p_0)/\partial\nu^2)\,(p_\delta - p_0) \cdot \nu \simeq \\ &\simeq \lambda - (\partial^2 U_\lambda(p_0)/\partial\tau^2)\,(\delta/\lambda) = \lambda - K_\lambda(p_0)\,\delta \end{aligned}$$

for small δ, where τ is the tangent vector. The assertion (3.2) follows from this and the fact that $|\nabla U_{\lambda,\delta}(p_\delta)| = |\nabla U_\lambda(p_\delta)|/(1-\delta)$. See [A6], §3 for the complete proof.

3.6. LEMMA. Let Γ_λ and Γ_α be solutions with $0 < \lambda \leq \alpha$. Assume that $0 < C \leq \lambda - K_\lambda(p) \leq M$ on Γ_λ (C, M constants) and that

$$\Gamma_{\lambda,s} \leq \Gamma_\alpha \leq \Gamma_{\lambda,t} \tag{3.5}$$

with $|s|, |t| \leq \delta_1(\lambda, C)$, where s is maximum and t is minimum subject to (3.5). Then $s \geq 0$, whence $\Gamma_\lambda \leq \Gamma_\alpha$. More generally, we have

$$((\alpha - \lambda)/2M) \leq s \leq t \leq (2(\alpha - \lambda)/C). \tag{3.6}$$

PROOF. For $p_s \in \Gamma_\alpha \cap \Gamma_{\lambda,s}$ and $p_t \in \Gamma_\alpha \cap \Gamma_{\lambda,t}$, we have

$$(3.7\text{a}, \text{b}) \qquad |\nabla U_{\lambda,t}(p_t)| \leq |\nabla U_\alpha(p_t)| = \alpha = |\nabla U_\alpha(p_s)| \leq |\nabla U_{\lambda,s}(p_s)|$$

as in the proof of Theorem 2.1. If $s < 0$, then (3.7 b) contradicts (3.4). Therefore, $0 \leq s \leq t$. It follows from (3.2) and «$\lambda - K_\lambda \leq M$ on Γ_λ» that $|\nabla U_{\lambda,s}(p_s)| \leq \lambda + 2Ms$, which, when combined with (3.7 b), implies that $(\alpha - \lambda) \leq 2Ms$. Similarly, it follows from (3.3) (with $\delta = t$) and (3.7 a) that $(\alpha - \lambda) \geq (Ct/2)$.

3.7. PROOF OF THEOREM 3.1. There exists a Lipschitz-continuous function $\phi(h)$ with $\phi(0) = 0$ such that if $h > 0$ is sufficiently small, then for each λ in the interval $I(h) = (\lambda_0 - h, \lambda_0 + h)$, there exists a solution Γ_λ satisfying $\Delta(\Gamma_\lambda, \Gamma_{\lambda_0}) < \phi(h)$. In fact this follows from Beurling's existence theorem (Theorem 3.2) and Lemma 3.5. By decreasing $h > 0$ if necessary, we can assume for some $\epsilon_0 > 0$ that U_λ is harmonic in $\Omega_{\lambda_0} \cup N_{\epsilon_0}(\Gamma_{\lambda_0})$ for all $\lambda \in I(h)$ (see Remark 3.3). Using this, and decreasing $h > 0$ again if necessary, one can show that Lemmas 3.5 and 3.6 apply in a uniform sense for all $\lambda, \alpha \in I(h)$ (i.e., we have $0 < C \leq \lambda - K_\lambda(p) \leq M$ on Γ_λ uniformly for all $\lambda \in I(h)$ and $\delta_1(\lambda, C) \geq \delta_1 > 0$ independent of $\lambda \in I(h)$). Thus Lemma 3.6 can be applied to every solution pair $\Gamma_\lambda, \Gamma_\alpha$ with $\lambda, \alpha \in I(h)$, showing that the solution family $\Gamma_\lambda, \lambda \in I(h)$ is elliptically ordered and Lipschitz- continuously varying. Finally, it is easy to apply this local result to obtain a maximal solution family with the same properties.

3.8. REMARK. Actually, our proof of Theorem 3.1 shows that for any compact subinterval J of I, there exists a constant $C > 0$ such that dist $(\Gamma_\alpha, \Gamma_\beta) \geq C|\alpha - \beta|$ uniformly for all $\alpha, \beta \in J$. This is due to the inequality (3.6) in Lemma 3.6.

4. INDIRECT VERIFICATION OF ELLIPTIC ORDERING

Any pratical application of the ideas in §3 requires that the condition «$K_\lambda(p) < \lambda$ on Γ_λ» be verifiable without actually computing the solution Γ_λ. In this section, we develop a pratical procedure, or test, by which the existence of large families of continuously-varying, elliptically-ordered solutions can be verified indirectly, without computing a single solution. In view of Theorems 2.1 and 3.1, this becomes a practical test for verifying the local uniqueness of solutions of Problem 1.1 in a given geometric situation. As in §3, we assume $n = 2$. We use $K^*(p)$ to denote the curvature of Γ^* at the point $p \in \Gamma^*$.

4.1. LEMMA. Given a solution Γ_λ, we have $K_\lambda(p) < \lambda$ on Γ_λ provided that either (a) $|\nabla U_\lambda| > \lambda/e$ on Γ^* or else (b) $K^*(p) < |\nabla U_\lambda|$ on Γ^*, where $e = \exp(1)$.

PROOF. The function $V_\lambda(p) = U_\lambda + \ln(|\nabla U_\lambda|/\lambda)$ is harmonic in Ω_λ and vanishes on Γ_λ. If condition (a) holds, then $V_\lambda(p) > 0$ on Γ^*. Then the maximum principle implies that $V_\lambda > 0$ in Ω_λ and $\partial V_\lambda/\partial\nu > 0$ on Γ_λ. The assertion follows, since $\partial V_\lambda/\partial\nu = \lambda - K_\lambda$ on Γ_λ. We also have $\partial V_\lambda/\partial\nu = |\nabla U_\lambda| - K^*(p)$ on Γ^*. Thus condition (b) also implies that $V_\lambda > 0$ in Ω_λ, from which the assertion follows as before.

4.2. A CONSTRUCTION. One chooses a C^2-curve $\Gamma \in \underline{X}$ such that

$$|\nabla U(p)| > K^*(p) \text{ on } \Gamma^*. \tag{4.1}$$

Subject to (4.1), one tries to choose Γ such that the number

$$M = \max\{|\nabla U(p)| : p \in \Gamma\}$$

is reasonably small.

4.3. THEOREM. Let $I = (M, \infty)$. Then there exists an elliptically-ordered, locally Lipshitz-continuously varying family of solutions $\Gamma_\lambda, \lambda \in I$, such that $\Gamma_\lambda > \Gamma$ for all $\lambda \in I$ and such that $\Gamma_\lambda \to \Gamma^*$ as $\lambda \to \infty$.

PROOF. For any $\lambda > M$ and solution $\Gamma_\lambda > \Gamma$, we have $|\nabla U_\lambda| > |\nabla U|$ on Γ^* by the maximum principle, so that $K_\lambda(p) < \lambda$ on Γ_λ by Lemma 4.1. Now choose $\lambda_0 > M$, let $\Gamma_{\lambda_0} > \Gamma$ be a solution fo Problem 1.1 at λ_0 (which exists by Theorem 3.2), and let $\Gamma_\lambda, a < \lambda < b$, denote the maximal solution family in Theorem 3.1, containing Γ_{λ_0} as a member. We claim that $b = \infty$. In fact if $b < \infty$, then as $\lambda \to b$ the solutions Γ_λ converge monotonically to a curve $\Gamma_b \in \underline{X}$, which can be shown to solve Problem 1.1 at the parameter value b. Since $\Gamma_b > \Gamma$, we have $K_b(p) < b$ on Γ_b, and it follows from Theorem 3.1 (with $\lambda_0 = b$) that the interval (a, b) is not maximal. Thus $b = \infty$, and it is quite easily shown using the monotonicity that $\Gamma_\lambda \to \Gamma^*$ as $\lambda \to \infty$. We also claim that $a \leq M$ and that $\Gamma_\lambda > \Gamma$ for all $\lambda > M$. In fact if either claim fails, then by Theorem 3.2 there exists a solution $\tilde{\Gamma}_{\lambda_1} > \Gamma$ of Problem 1.1 at $\lambda_1 > M$ such that $\tilde{\Gamma}_{\lambda_1}$ is not a member of the original maximal family. Starting with $\tilde{\Gamma}_{\lambda_1}$, Theorem 3.1 generates a second maximal solution family $\tilde{\Gamma}_{\lambda_1}, \tilde{a} < \lambda < \infty$, such that $\tilde{\Gamma}_{\lambda_1} \to \Gamma^*$ as $\lambda \to \infty$. But then Corollary 2.2 implies that $\tilde{\Gamma}_\lambda = \Gamma_\lambda$ for all $\lambda > \max(a, \tilde{a})$, and since both families are maximal, we have $\tilde{a} = a$. Thus $\tilde{\Gamma}_{\lambda_1}$ is a member of the original maximal family, a contradiction which proves our assertions.

4.4. REMARK. For any $\lambda > M$ (in the context of Theorem 4.3), Γ_λ is the unique solution of Problem 1.1 at λ such that $\Gamma_\lambda > \Gamma$.

5. CONVERGENCE RESULTS FOR A TRIAL FREE-BOUNDARY METHOD

In a typical free-boundary problem in PDE's, the free boundary must be chosen such that a certain overspecified boundary value problem (involving both Dirichlet and Neumann boundary conditions on the free boundary) can be solved. A natural method of successive approximations for computing the free boundary is called the trial free-boundary method. One step of this iterative procedure consists of starting with a curve (or surface) Γ_k called the kth iterate, solving a boundary value problem in the resulting domain Ω_k, which involves either the Dirichlet condition or the Neumann condition or some combination of both on Γ_k, and using the solution U_k of this problem to construct a new boundary curve (or surface) Γ_{k+1} which comes closer to satisfying both boundary conditions. For detailed discussions of this general method, see Crank [C], Chapter 8, and Cryer [Cr]. Although this very natural method for computing free boundaries has been used for more than 70 years, and is frequently used at present, there appear to be no convergence proofs in the literature, except for the author's results in [A4], and [A5], which apply to the exterior Bernoulli free-boundary problem in the starlike case (including the case where the star center is at $y = -\infty$, causing the relevant curves to be graphs of functions of x). Actually, it must be realized that even after the free-boundary problem has been specified, there will usually be several distinct ways in which the trial free-boundary iteration (from Γ_k to Γ_{k+1}) can be reasonably defined, each leading to its own convergence behavior. Our purpose in this section is to show that for the author's particular trial free-boundary method (from [A4], [A5]), transferred to the context of Problem 1.1 in 2-dimensions, the convergence behaviour of the iterates is closely related to the elliptic-ordering property of the solutions Γ_λ.

5.1. DEFINITIONS. (See Fig. 3). In Problem 1.1 (with $n = 2$), let a point $p_0 \in G$ be specified, and let $\underline{\overline{X}}(p_0) = \{\Gamma \in \underline{\overline{X}} : p_0 \in D\}$. Given $\Gamma \in \underline{\overline{X}}(p_0)$ and $\lambda > 0$, we define

$$T_{\lambda,\epsilon}(\Gamma) = \Psi_{\epsilon/\lambda}(\Phi_\epsilon((\Gamma)) \in \underline{\overline{X}}(p_0) \tag{5.1}$$

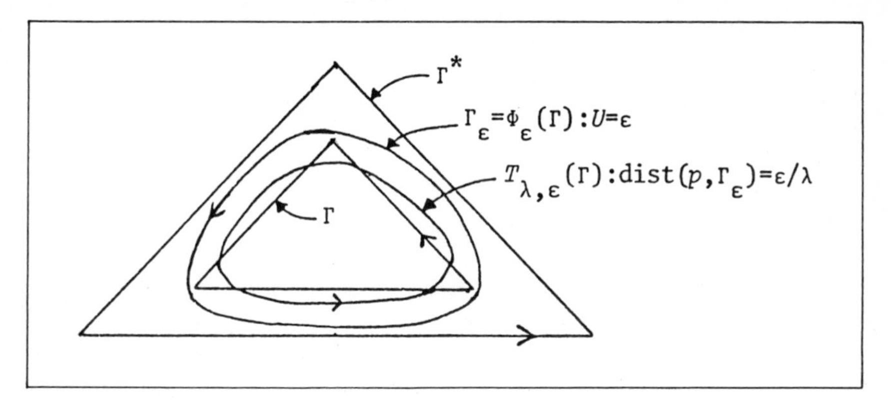

Fig. 3.

for sufficiently small $\epsilon > 0$. Here

$$\Phi_\epsilon(\Gamma) = \{p \in \Omega : U(p) = \epsilon\} \in \underline{X}(p_0) \tag{5.2}$$

for $0 < \epsilon < 1$ and $\Psi_\delta(\Gamma) = \partial E_\delta$ for $0 < \delta < \text{dist}\,(p_0, \Gamma)$, where E_δ denotes the maximal connected subset of D such that $p_0 \in E_\delta$ and such that dist $(p, \Gamma) > \delta$ for all points $p \in E_\delta$. Notice that $T_{\lambda,\epsilon}(\Gamma)$ is defined whenever $0 < \epsilon < \min(1, \lambda\text{dist}\,(p_0, \Gamma))$.

5.2. REMARK. The normal variation (at p) of a sufficiently-regular curve $\Gamma \in \underline{X}(p_0)$ resulting from the application of the operator $T_{\lambda,\epsilon}$ (for small $\epsilon > 0$) is approximately

$$\delta\nu(p) \simeq ((1/|\nabla U|) - (1/\lambda))\epsilon. \tag{5.3}$$

Therefore, if $|\nabla U| < \alpha < \lambda$ (resp. $|\nabla U| > \alpha > \lambda$) on Γ, then

$$\delta\nu(p) > (<)\, (\lambda - \alpha)\, (\epsilon/\lambda\alpha) > (<) 0 \tag{5.4}$$

on Γ for sufficiently small $\epsilon > 0$.

5.3. LEMMA. The operators $T_{\lambda,\epsilon}(\Gamma)$ are monotone in the following sense: we have $T_{\lambda,\epsilon}(\Gamma_1) \leq T_{\lambda,\epsilon}(\Gamma_2)$ for any $\lambda > 0, 0 < \epsilon < 1$, and curves $\Gamma_1, \Gamma_2 \in \underline{X}(p_0)$ such that $\Gamma_1 \leq \Gamma_2$ and $0 < \epsilon < \lambda\,\text{dist}\,(p_0, \Gamma_1)$.

PROOF SKETCH. The monotonocity of $\Phi_\epsilon : \underline{X} \to \underline{X}$ follows from the maximum principle, and the monotonocity of Ψ_δ follows from elementary properties of the distance function.

5.4. REMARK. For the following two theorems, let $\Gamma_\lambda, \lambda \in J = [a,b]$, denote an elliptically-ordered, Lipschitz-continuously varying solution family, and let $\tilde{\underline{\overline{X}}} = \{\Gamma \in \underline{\overline{X}}(p_0) : \Gamma_a \leq \Gamma \leq \Gamma_b\}$, where we choose $p_0 \in D_a$ in the definition of $\underline{\overline{X}}(p_0)$. Given $\lambda_0 \in (a,b)$, it is easily seen using Remark 5.2 (applied to Γ_a, Γ_b) and Lemma 5.3 that if $\rho_0 > 0$ is sufficiently small, then $T_{\lambda_0,\epsilon} : \tilde{\underline{\overline{X}}} \to \tilde{\underline{\overline{X}}}$ for all values $0 < \epsilon \leq \rho_0$.

5.5. THEOREM. In the context of Remark 5.4, there exist constants $A, B > 0$ such that for any $\Gamma \in \tilde{\underline{\overline{X}}}$ and any constants $\alpha, \beta \in J$ such that $\alpha \leq \lambda_0 \leq \beta$ and $\Gamma_\alpha \leq \Gamma \leq \Gamma_\beta$, we have

$$\Gamma_{\alpha(\epsilon)} \leq T_{\lambda_0,\epsilon}(\Gamma_\alpha) \leq T_{\lambda_0,\epsilon}(\Gamma) \leq T_{\lambda_0,\epsilon}(\Gamma_\beta) \leq \Gamma_{\beta(\epsilon)} \tag{5.5}$$

for $0 < \epsilon < \epsilon_0$, where

$$\lambda_0 - \alpha(\epsilon) \leq (1 - A\epsilon)(\lambda_0 - \alpha) + B\epsilon^2 \tag{5.6}$$

$$\beta(\epsilon) - \lambda_0 \leq (1 - A\epsilon)(\beta - \lambda_0) + B\epsilon^2. \tag{5.7}$$

PROOF SKETCH. The estimate (5.6) and (5.7) follow essentially from (5.4) and the Lipschitz-continuous dependence of Γ_λ on λ. The middle two inequalities in (5.5) are due to Lemma 5.3. For the complete proof, see [A6], Theorem 5.4.

5.6. THEOREM. In the context of Remark 5.4, let the sequence of curves $\gamma_1, \gamma_2, \gamma_3, \ldots$, all in $\tilde{\underline{\overline{X}}}$, be defined recursively such that

$$\gamma_{k+1} = T_{\lambda_0,\epsilon_k}(\gamma_k), k = 1, 2, 3 \ldots,$$

where $\gamma_1 = \Gamma \in \tilde{\underline{\overline{X}}}$ and $\epsilon_1, \epsilon_2, \epsilon_3, \ldots$ is an arbitrary positive nullsequence with divergent sum, such that the individual terms all satisfy $\epsilon_k \leq \rho_0$. Then $\gamma_k \to \Gamma_{\lambda_0}$ as $k \to \infty$ in the sense that $\Delta(\gamma_k, \Gamma_{\lambda_0}) \to 0$.

PROOF SKETCH. For the kth iterate γ_k, choose values $\alpha_k, \beta_k \in J$ with α_k maximum and β_k minimum subject to the requirements that $\alpha_k \leq \lambda_0 \leq \beta_k$ and $\Gamma_{\alpha_k} \leq \gamma_k \leq \Gamma_{\beta_k}$. Then the estimates (5.6), (5.7) imply recursion relations for the sequences (α_k) and (β_k) which show $\alpha_k, \beta_k \to \lambda_0$ as $k \to \infty$. For the details, see [A5], §4.

5.7. REMARK. If the sequence (ϵ_k) in Theorem 5.6 is chosen suitably (essentially $\epsilon_k \sim 1/k$), then one obtains $\Delta(\gamma_k, \Gamma_{\lambda_0}) = O(1/k)$ as $k \to \infty$ (for proof, see [A5], Remark 4). Examples show that in general, convergence for this method cannot be faster than $O(1/k)$.

5.8. REMARK. The author has recently extended the convergence proof in [A5] to an analogue trial free-boundary iteration for the 2-phase fluid problem in the n-dimensional, starlike case. See [A9].

6. BEURLING'S THEOREM IN THE n-DIMENSIONAL, CONVEX CASE

The key to generalizing Theorem 3.1 to the interior Bernoulli free-boundary problem in n-dimensions is a corresponding generalization of the Beurling existence result (Theorem 3.2). However, such a generalization appears difficult to obtain in the arbitrary geometrical situation. Therefore, we restrict attention to the convex case, in which G is convex and we seek convex solutions. We define $\underline{X}_C = \{\Gamma \in \underline{X} : D \text{ is convex}\}$.

6.1. THEOREM. Assume in Problem 1.1 that $n \geq 3$ and G is convex. Given $\lambda > 0$, let $\Gamma_{\pm} \in \underline{X}$ be two C^2-hypersurfaces such that (a) $\Gamma_- < \Gamma < \Gamma_+$ for at least one convex hypersurface $\Gamma \in \underline{X}_C$ and (b) $|\nabla U_+| > \lambda$ on Γ_+ and $|\nabla U_-| < \lambda$ on Γ_-. Then there exists a convex solution $\Gamma_\lambda \in \underline{X}_C$ of Problem 1.1 (at λ) such that $\Gamma_- < \Gamma_\lambda < \Gamma_+$.

The remainder of this section is devoted to the proof of Theorem 6.1. This proof, which uses ideas developed by the author in [A2], [A3] (see also [C-S] and [F]), is based on the operators $T_{\lambda,\epsilon}(\Gamma) = \Psi_{\epsilon/\lambda}(\Phi_\epsilon(\Gamma))$ defined in §5 (see (5.1), (5.2)), the one distinction being that in the convex case $(\Gamma \in \underline{X}_C)$ it sufficies to define

$$\Psi_\delta(\Gamma) = \{p \in D : \text{dist}\,(p, \Gamma) = \delta\} \in \underline{X}_C \tag{6.1}$$

for $\delta > 0$.

6.2. LEMMA. Given bounded, convex nested domains G and D in $\mathbf{R}^n$ with $\bar{D} \subset G$, let U denote the capacity potential in $\Omega := G \backslash \bar{D}$. Then (a) all the level hypersurfaces of U are convex, and (b) the function $\phi(p) = \ln(|\nabla U|)$ is superharmonic in Ω.

PROOF. For the proof of Part (a), see [C-S] or [Ko] (also [G], [K2], [Ke]). For

Part (b), one shows using $\Delta U = 0$ that

$$\Delta \phi(p_0) = \sum_{i,j=1}^{n-1} \left(S_{ij}^2 - S_{ii}S_{jj} \right),$$

where $S_{ij} = \partial^2 S/\partial x_i \partial x_j$ and where $x_n = S(x_1, x_2, \dots, x_{n-1})$ is a local parametrization of the level hypersurface of U through p_0 such that $S_i = 0, i = 1, \dots, n-1$, at p_0. Since the local parametrization has an extremum at p_0, due to the convexity of the hypersurface, the 2-dimensional discriminants $(S_{ii}S_{jj} - S_{ij}^2)$ are all non-negative at p_0.

6.3. REMARK. Due to Lemma 6.2 (a), the operators $T_{\lambda,\epsilon}(\Gamma)$ defined by (5.2), (6.1), (5.1) preserve convexity. We have $T_{\lambda,\epsilon} : \underline{\overline{X}}_C \to \underline{\overline{X}}_C$ for $\lambda > 0, 0 < \epsilon < 1$, where $\underline{\overline{X}}_C$ includes the empty set.

6.4. DEFINITION. We define the functional $f(\Gamma) : \underline{\overline{X}}_C \to \mathbf{R}$ such that

$$f(\Gamma) = K(\Omega) + \lambda^2|\Omega| = \int_\Omega (|\nabla U|^2 + \lambda^2) \, d\,x, \tag{6.2}$$

where $K(\Omega)$ and $|\Omega|$ denote the capacity and the n-dimensional volume of Ω.

6.5. LEMMA. Given $\Gamma \in \underline{\overline{X}}_C$ and $\lambda > 0$, let $\Gamma_\epsilon = T_{\lambda,\epsilon}(\Gamma) \in \underline{\overline{X}}_C$ for small $\epsilon > 0$. Then

$$f(\Gamma_\epsilon) \le f(\Gamma) - \epsilon P(\Gamma; \epsilon) + O(\epsilon^2) \tag{6.3}$$

as $\epsilon \to 0+$, where

$$P(\Gamma; \epsilon) = \frac{1}{\epsilon} \int_0^\epsilon \int_{\Gamma_\alpha} \left(\frac{(|\nabla U| - \lambda)^2}{|\nabla U|} \right) d\,\sigma \, d\,\alpha. \tag{6.4}$$

Here $\Gamma_\alpha = \Phi_\alpha(\Gamma)$ and $d\,\sigma$ denotes the differential Euclidean area of the hypersurface.

PROOF SKETCH. Assume that Γ is a sufficiently regular hypersurface, so that the well-known variational formulas for capacity and volume can be applied. Using (5.3), we see that the variation in $f(\Gamma)$ caused by the operator $T_{\lambda,\epsilon}(\Gamma)$ is given approximately (for small $\epsilon > 0$) by

$$\begin{aligned} \delta f &\simeq \int_\Gamma (|\nabla U|^2 - \lambda^2) \delta\nu \, d\,\sigma \simeq \int_\Gamma (|\nabla U|^2 - \lambda^2) \left(\frac{1}{|\nabla U|} - \frac{1}{\lambda} \right) \epsilon \, d\,\sigma \\ &\simeq -\epsilon \int_\Gamma \left(\frac{|\nabla U| + \lambda}{\lambda |\nabla U|} \right) (|\nabla U| - \lambda)^2 \, d\,\sigma. \end{aligned}$$

The rigorous proof is given in [A7].

PROOF SKETCH FOR THEOREM 6.1. Given $\Gamma_\pm \in \overline{X}$, let $\tilde{\overline{X}}_C = \{\Gamma \in \overline{X}_C : \Gamma_- \leq \Gamma \leq \Gamma_+\}$, which is non-empty due to assumption (a). We have $T_{\lambda,\epsilon} : \tilde{\overline{X}}_C \to \tilde{\overline{X}}_C$ for sufficiently small $\epsilon > 0$, due to assumption (b), the estimate (5.3), and the monotonicity of $T_{\lambda,\epsilon}$ (see Lemma 5.3). Choose Γ_λ to be a minimizer of the functional $f(\Gamma) : \tilde{\overline{X}}_C \to \mathbb{R}$ defined by (6.2). Clearly $f(T_{\lambda,\epsilon}(\Gamma_\lambda)) \geq f(\Gamma_\lambda)$. Using Lemma 6.5, we conclude that $P(\Gamma_\lambda; \epsilon) = O(\epsilon)$ as $\epsilon \to 0+$. Thus, there exists a positive null-sequence (ϵ_i) such that

$$\int_{\Gamma_{\epsilon_i}} ((|\nabla U_\lambda| - \lambda)^2 / |\nabla U_\lambda|) \, d\sigma = O(\epsilon_i)$$

as $i \to \infty$, where $\Gamma_{\epsilon_i} = \Phi_{\epsilon_i}(\Gamma_\lambda)$. Thus Γ_λ satisfies the free-boundary condition (1.1) in a weak sense, and one can go on to show that Γ_λ is a classical solution of Problem 1.1 (See [A2], [A7]).

7. EXISTENCE OF ELLIPTICALLY-ORDERED SOLUTION FAMILIES IN THE n-DIMENSIONAL, CONVEX CASE

The results in §2 are n-dimensional, and all the 2-dimensional results §§3, 4, and 5 can be adapted to Problem 1.1 in the n-dimensional, convex case (where G is convex and one seeks convex solutions Γ_λ) by making a few small changes. Here, we discuss these changes. To begin with, in n-dimensions, one must redefine $K_\lambda(p)$ to denote $(n-1)$ times the mean curvature of the $(n-1)$-dimensional hypersurface $\Gamma_\lambda \in \overline{X}_C$ at $p \in \Gamma_\lambda$. We have $K_\lambda(p) \geq 0$ on Γ_λ due to the convexity of D_λ. (The n-dimensional definition of $K^*(p)$ on Γ^* is analogous.) With this definition of $K_\lambda(p)$, Theorem 3.1 carries over without change except that we must now assume that $\Gamma_{\lambda_0} \in \overline{X}_C$ and the theorem now asserts that $\Gamma_\lambda \in \overline{X}_C$ for all members of the maximal, elliptically ordered solution family. The proof follows from Theorem 6.1 in conjunction with Lemmas 3.5 and 3.6, which carry over to n dimensions without change. Turning to §4, the main obstacle to generalizing these results to n-dimensions is the fact that $\ln(|\nabla U_\lambda|)$ is not a harmonic function for $n \geq 3$. However, this difficulty is overcome in the convex case by the second part of Lemma 6.2., which shows that $\ln(|\nabla U_\lambda|)$ is superharmonic in Ω_λ. Using this, Lemma 4.1 carries over without change when $\Gamma_\lambda \in \overline{X}_C$. Thus, assuming that $\Gamma \in \overline{X}_C$ in Construction 4.2, Theorem 4.3 and its proof carry over without change, except that all the solutions $\Gamma_\lambda, \lambda > M$, are in $\overline{X}_C$. By combining the modified Theorem 4.3 with Theorem 2.1, we see that for each $\lambda > M, \Gamma_\lambda$ is the unique convex solution of Problem 1.1 at λ, such that $\Gamma_\lambda > \Gamma$. Turning to §5, since the operators $T_{\lambda,\epsilon}$ preserve convexity, the entire discussion in §5 can be restricted to the convex case, where it generalizes to n-dimensions with no change in the proofs. Thus, we have obtained the following theorem:

7.1. THEOREM. Assume in Problem 1.1 that $n \geq 3$ and G is convex. Choose a convex C^2-hypersurface $\Gamma \in \overline{X}_C$ such that $|\nabla U| > K^*$ on Γ^*, and let $I = (M, \infty)$, where $|\nabla U(p)| \leq M$ on Γ. Then there exists an elliptically-ordered, locally Lipschitz-continuously varying solution family $\Gamma_\lambda, \lambda \in I$, such that $\Gamma_\lambda > \Gamma$ and $\Gamma_\lambda \in \overline{X}_C$ both for all $\lambda \in I$, and such that $\Gamma_\lambda \to \Gamma^*$ as $\lambda \to \infty$. Given $\lambda_0 \in I, \Gamma_{\lambda_0}$ is the unique convex solution of Problem 1.1 at λ_0 subject to the side condition that $\Gamma_{\lambda_0} > \Gamma$. Moreover, let (γ_k) be any sequence of convex hypersurfaces defined recursively by $\gamma_{k+1} = T_{\lambda_0, \epsilon_k}(\gamma_k), k = 1, 2, 3, \ldots$, where $(\gamma_1 \in \overline{X}_C$ and $\gamma_1 > \Gamma_\lambda$ for some $\lambda \in I$, and where (ϵ_k) is a positive nullsequence with divergent sum such that $0 < \epsilon_k < r$ for all k. Then $\gamma_k \to \Gamma_{\lambda_0}$ as $k \to \infty$ provided that $r > 0$ is sufficiently small.

REFERENCES

[A1] A. ACKER: Heat-flow inequalities with applications to heat-flow optimization problems. *SIAM J. Math. Anal.* **8** (1977), 604-618.

[A2] A. ACKER: Interior free-boundary problems for the Laplace equation. Arc. Rat'l. *Mech. Anal.* **75** (1981), 157-168.

[A3] A. ACKER: On the convexity of equilibrium plasma configurations. *Math. Meth. Appl. Sci.* **3** (1981), 435-443.

[A4] A. ACKER: How to approximate the solutions of certain free-boundary problems for the Laplace equation by using the contraction principle. *J. Appl. Math. Phys. (ZAMP)* **32** (1981), 22-33.

[A5] A. ACKER: Convergence results for an analytical trial free-boundary method. *IMA J. of Numerical Analysis* **8** (1988), 357-364.

[A6] A. ACKER: On the qualitative theory of parametrized families of free boundaries. J. reine angew. Math. **393** (1989), 134-167. Preliminary Version: Preprint no. 422, Sonderforschungbereich 123, Heidelberg University, June, 1987.

[A7] A. ACKER: Uniqueness and monotonicity of solutions for the interior Bernoulli free boundary problem in the convex, n-dimensional case. Nonlinear Analysis (TMA) (to appear).

[A8] A. ACKER: On the non-convexity of solutions in free-boundary problems arising in plasma physics and fluid dynamics . *Comm. Pure Appl. Math.* (to appear).

[A9] A. ACKER: On the convexity and successive approximation of solutions in a Bernoulli free-boundary problem with two fluid phases. Comm. in PDE (to appear).

[A-C] H. W. ALT and L. A. CAFFARELLI: Existence and regularity for minimum problems with free boundary. *J. reine angew. Math.* **325** (1981), 105-144.

[B] A. BEURLING: On free boundary problems for the Laplace equation. *Sem. on Analytic functions* **1** (1957), Inst. for Advanced Study, Princeton, N. J., 248-263.

[B-Z] G. BIRKHOFF and E. H. ZARANTONELLO: *Jets, Wakes, and Cavities*, Academic Press, 1957.

[C-S] L. A. CAFFARELLI and J. SPRUCK: Convexity properties of solutions to some classical variational problems. *Comm. in P. D. E.* **7** (1982), 1337-1379.

[C] J. CRANK: *Free and Moving Boundary Problems.* Clarendon Press, Oxford, 1984.

[Cr] C. W. CRYER: Numerical methods for free and moving boundary problems. Document No. 2/86-N., Inst für numerische und instrumentelle Mathematik, Westfälische Wilhelms-Universität zu Münster, Einsteinstr. 62; D-4000 Münster, West Germany.

[D] I. I. DANILJUK: On integral Functionals with a Variable Domain of Integration. *Proc. Steklov Inst. of Math.* **118** (1972). English: AMS (1976).

[F] A. FRIEDMAN: *Variational Principles and Free-Boundary Problems.* New York: John Wiley and Sons (1982).

[G] R. GABRIEL: A result concerning convex level surfaces of 3-dimensional harmonic functions. *J. London Math. Soc.* **32** (1957), 286-294.

[Ga] D. GAIER: On an area problem in conformal mapping. *Results in Mathematics* **10** (1986), 66-81.

[K1] B. KAWOHL: *Rearrangements and Convexity of Level Sets in PDE.* New York: Springer-Verlag, 1985.

[K2] B. KAWOHL: When are the solutions to nonlinear elliptic boundary value problems convex? *Comm. in P.D.E.* **10** (1985), 1223-1225.

[Ke] A. KENNINGTON: Power concavity and boundary value problems. *Indiana U. Math. J.* **34** (1985), 687-704.

[Ko] N. KOREVAAR: Convex solutions of nonlinear elliptic and parabolic boundary value problems. *Indiana U. Math. J.* **32** (1983), 603-614.

[L-S] M. A. LAVRENTIEV and B. W. SHABAT: *Methoden der Komplexen Funktionentheorie.* Berlin: VEB Deutscher Verlag der Wissenshaften, 1967.

[Le] H. LEWY: A note on harmonic functions and a hydrodynamical application. *Proc. Amer. Math. Soc.* **3** (1952), 111-113.

[P-W] M. H. PROTTER and H. F. WEINBERGER: *Maximum Principles in Differential Equations.* New York Springer-Verlag, 1984 (originally published in 1967).

[T] D. E. TEPPER: Free-boundary problem-the starlike case. *SIAM J. Math. Anal.* **6** (1975), 503-505.

[W] W. WALTER: *Differential and Integral Inequalities.* New York: Springer-Verlag, 1970.

A BLOW UP PHENOMENON FOR REACTION DIFFUSION EQUATIONS IN UNBOUNDED DOMAINS

CATHERINE BANDLE

1. THE MODEL CASE

Let $\mathcal{D} \subseteq \mathbf{R}^N$ be an arbitrary domain, let $x = \{x_1, \ldots, x_N)$ denote a generic point and $r := |x|$ be its distance from the origin. We shall consider reaction-diffusion problems of the type

$$(P) \qquad \begin{cases} u_t - \Delta u &= t^q r^\sigma u^p & \text{in} & \mathcal{D} \times (0,T) \\ u &= 0 & \text{in} & \partial\mathcal{D} \times (0,T) \\ u(x,0) &= u_0(x) \geq 0 & & \\ e^{-cr^2} u(x,t) &\to 0 & \text{as} & |x| \to \infty \forall c > 0 \end{cases}$$

where $q > 0, p > 1$ are arbitrary real numbers and $\sigma \in \mathbf{R}$ is such that (P) admits a unique local solution in the classical sense. In view of the growth condition for large $|x|$, the parabolic maximum principle applies. It implies that the solution of (P) is strictly positive in $\mathcal{D}$ for any $t > 0$ unless $u_0(x) \equiv 0$. Thus concerning the behaviour of $u(x,t)$ as a function of the time t, only two possibilities can occur:

(i) $u(x,t)$ exists for all $t > 0 [T = \infty]$ (in this case u called a *global* solution)

(ii) $u(x,t)$ *blows up* in finite time, i.e. $\limsup_{t \nearrow T} \sup_{x \in D} u(x,t) = \infty$ for some $T < \infty$

In the linear case $p = 1$ all solutions are global. If $p > 1$, not all solutions are global, an observation due to Kaplan [Kap] for the particular nonlinearity $q = \sigma = 0$.

Physically $u(x,t)$ can be conceived as the non-dimensional temperature of a chemical reactant occupying a region $\mathcal{D}$, and kept at a constant value zero on the boundary $\partial\mathcal{D}$. In this greatly idealized model, $f(t,x,u) := t^q r^\sigma u^p$ describes the heat source.

2. FUJITA'S RESULT

In a classical paper of 1966, Fujita [F1] proved for the special case $\mathcal{D} = \mathbf{R}^N$, $q = \sigma = 0$ the following surprising result:

(i) If $1 < p < 1 + 2/N$ every non-trivial solution blows up in finite time.

(ii) If $p > 1 + 2/N$ there exist global solutions for sufficiently small initial data.

Later several authors [H, AW, W] established that for the cutoff value $p^* = 1 + 2/N$ no non-trivial global solutions are possible. Fujita himself and several other authors [F2, M1] extended this result to problems with more general nonlinearities.

It should be noticed that for $p < 1$ blow up never occurs. In fact the function $z(t)$ satisfying $z_t = z^p, z(0) = |u_0|_\infty (:= \sup_{\mathcal{D}} |u_0(x)|)$ is an upper solution and remains bounded if $p < 1$.

The difference between (i) and (ii) is due to the fact that for large p and small initial data the heat source isn't strong enough to cause a considerable increase in temperature.

Problem (P) with $\mathcal{D} = \mathbf{R}^N$ has been studied subsequently in [M1] for $\sigma = 0$, in [Kav, BL1] for $q = 0$ and [B, BL2] for general σ and q. The final result is:

THEOREM 2.1. *Let $\mathcal{D} = \mathbf{R}^N$ and assume $\sigma > -2$ in order to guarantee local existence.*

(i) If $1 < p < 1 + \dfrac{2 + 2q + \sigma}{N} =: p^*$ *no global non-trivial solutions of (P) exist.*

(ii) For $p > p^*$ global solutions do exist.

Recently Levine and Meier have shown that (i) holds also for p^*.

DEFINITION. Problem (P) is said to have the *property (F)*, if all non-trivial solutions blow up in finite time.

It is easily seen (cf. e.g. [B, M2] that (P) cannot have the property (F) if $\mathcal{D}$ is bounded. Consequently the same is true for cylindrical domains $\mathcal{D} = \mathcal{D}^0 \times \mathbf{R}^k$ where $\mathcal{D}^0 \subset \mathbf{R}^{N-K}$ is bounded.

The aim of this survey is to give examples of problems having the property (F), and to discuss some related topics such as the large time behaviour of global solutions and existence and nonexistence of stationary states.

For the maximum principle it follows immediatly:

MONOTONICITY LEMMA 2.2. *If* (P) *has the property* (F), *then the problem* $(P)'$ *corresponding to problem* (P) *with* $\mathcal{D}$ *replaced by* $\mathcal{D}' \supset \mathcal{D}$ *has also this property.*

Recently Meier [M2] discovered a relation between (F) and the asymptotic behaviour of the solutions of the linear heat equation for large t.

LEMMA 2.3. *Assume* $\sigma = 0$ *and let* $w(x,t) \not\equiv 0$ *denote a positive solution of the linear equation* $w_t = \Delta w = 0$ *in* $\mathcal{D} \times \mathbb{R}^+, w = 0$ *in* $\partial\mathcal{D} \times \mathbb{R}^+$. *Put* $s^* = \sup\{s$ *such that* $\lim_{t\to\infty} t^s|w|_\infty < \infty$ *for some* $w \not\equiv 0\}$.

(i) For all $p \in (1, p^*), p^* = 1 + (1+q)/s^*, (P)$ *has the property* (F)

(ii) Property (F) *cannot hold if* $p > p^*$.

Here again the discussion of the cutoff case seems to be much more delicate. All examples indicate that the situation should be similar if $\sigma \neq 0$ in the sense that there is a critical $p^* = p^*(q, \sigma, \mathcal{D})$ for which the statements (i) and (ii) hold.

3. EXTERIOR DOMAINS

Let $\mathcal{D}_i$ be a collection of bounded, simply connected domains, contained in same sphere of radius $R < \infty$. Set $D = \mathbb{R}^N - \bar{\mathcal{D}}_i$. (cf. Fig. 1.)

In this case we have [B, BL1, BL2].

THEOREM 3.1. *The same statements hold as inTheorem 2.1 for the whole space.*

This theorem leads to the conjecture that «small» changes of an infinite $\mathcal{D}$ in a finite region don't affect existence and non-existence of global solutions. This is certainly not correct if $\mathcal{D}$ is bounded.

The next lemma is based on the maximum principle.

LEMMA 3.2. *Let* $\mathcal{D}$ *be fixed. If a reaction-diffusion equation with a heat source* $f(t,x,u)$ *has the property* (F), *then the same is true for every other problem with a heat source* $\tilde{f}(t,x,u) \geq f(t,x,u)$ *in* $\mathbb{R}^+ \times \mathcal{D} \times \mathbb{R}^+$.

The monotonicity Lemmas 2.2 and 3.2 together with Theorem 3.1 allow us to treat problems with nonlinearities of the form $k(x,t)u^p, k(x,t)$ being a non-negative function subject to the condition

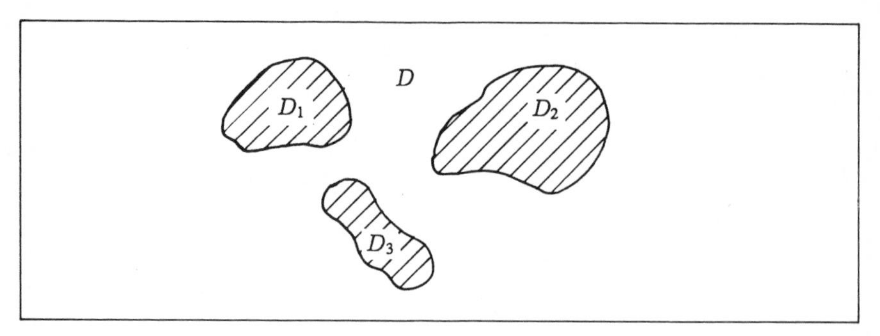

Fig. 1.

$(K)_+$ $\qquad k(x,t) \geq t^q r^\sigma$ for $t \geq t_0$ and $|x| \geq R$, $(t_0, R > 0$ arbitrary$)$

COROLLARY 3.3 *Let $\mathcal{D}$ be the whole space or an exterior domain and consider (P) with a nonlinearity $k(t,x)u^p$, where k satisfies $(K)_+$. If $p \in (1, 1+(2+2q+\sigma)/N)$, then (P) has the property (F).*

4. CONE LIKE DOMAINS

In order to describe a cone like domain, let us introduce the polar coordinates (r,θ) of a point x; θ stands for the angles $(\theta_1, \dots, \theta_{N-2}, \varphi)$ where $0 \leq \theta_i \leq \pi, i = 1, \dots, N-2$ and $0 \leq \varphi < 2\pi$, and r is the distance of x from the origin. Let $\Omega \subseteq S^{N-1} := \{x : |x| = 1\}$ be an arbitrary connected domain. Then

$$\mathcal{D} = \{re : e \in \Omega, r > 0\} \tag{4.1}$$

is called a *cone like domain.* In the polar coordinates the Laplace operator has the form

$$\Delta = \Delta_r + \frac{1}{r^2}\Delta_\theta, \tag{4.2}$$

where $\Delta_r = \dfrac{1}{r^{N-1}} \dfrac{\partial}{\partial r}\left(r^{N-1}\dfrac{\partial}{\partial r}\right)$ is the radial part and Δ_θ is the Beltrami operator on the sphere S^{N-1}.

Let ω be the lowest eigenvalue of

$$\Delta_\theta \psi + \omega\psi = 0 \text{ in } \Omega, \psi = 0 \text{ in } \partial\Omega \tag{4.3}$$

and write $\gamma_+(\gamma_-)$ for the positive (negative) root of

$$(4.4) \qquad \gamma^2 + \gamma(N-2) - \omega = 0.$$

As in the previous cases there is an interval $(1, p^*)$ of p-values for which (P) has the property (F) whereas (F) doensn't hold for $p > p^*$. More precisely:

THEOREM 4.1. *Let* $p^* := 1 + (2 + 2q + \sigma)/(2 - \gamma_-)$.
(i) If $p \leq p^*, (P)$ *has the property* (F).
(ii) For $p > p^*$, *property* (F) *cannot persist.*

A special case of part (i) was first established by Meier [M1]. The result stated here follows immediatly from [BL1] and [B, BL2]. The second assertion is due to Levine and Meier [LM]. A different approach was carried out by Kavian [Kav], suitable to solutions satisfying a strong decay at infinity. Kavian expresses, in the case $q = \sigma = 0, p^*$ in terms of the principal eigenvalue of a certain eigenvalue problem defined in $\mathcal{D}$.

REMARKS. (i) The theorem remains true if $\mathcal{D}$ is replaced by an «exterior cone» $\mathcal{D}_R = \mathcal{D} \cup \{|x| > R\}$. Therefore if we consider problems with a more general nonlinearity $k(t, x)u^p$, where $k(x, t)$ satisfies $(K)_+$, then the property (F) continues to hold for $p \in (1, 1 + (2 + 2q + \sigma)/(2 - \gamma_-))$.

(ii) For the sector $\mathcal{D} = \{(r, \varphi) : 0 \leq \varphi < \pi\nu\}$ in the plane whe have $\omega_1 = 1/\nu^2$ and thus $p^* = 1 + \nu(2 + 2q + \sigma)/(2\nu + 1)$. Observe that for the domain $\mathcal{D} - \{x > 0\}$ (cf. Fig. 2.) we get $p^* = (9 + 4q + 2\sigma)/5$, which is for $\sigma = q = 0$, smaller than the corresponding value $p^* = \dfrac{4 + 4q + 2\sigma}{2}$ for the whole plane.

An interesting question raised by I. Stakgold is: What is the critical value p^* for a slit domain $\mathcal{D}' = \mathbb{R}^2 - \sum_{i=1}^{k} \{(r, \theta_i) : r \geq r_i > 0\}$ (cf. Fig. 3)? From the monotonicity Lemma 2.2 we deduce that $p^* \geq 1 + \nu_0(2 + 2q + \sigma)/(2\nu_0 + 1) =: \tilde{p}$ where $\pi\nu_0$ is the smallest angle between the rays $\theta = \theta_i$. Note that if $u(x, t)$ is a solution of $\mathcal{D}'$ then for arbitrary $\beta > 0$

$$v(x, t) = \beta^{(2+2q+\sigma)/(p-1)} u(\beta x, \beta^2 t)$$

is a solution in $\mathcal{D}'_\beta = \mathbb{R}^2 - \sum_{i=1}^{k} \{(r, \theta_i) : r \geq r_i/\beta\}$. This observation together with remark (i) leads to the conjecture that $p^* = \tilde{p}$.

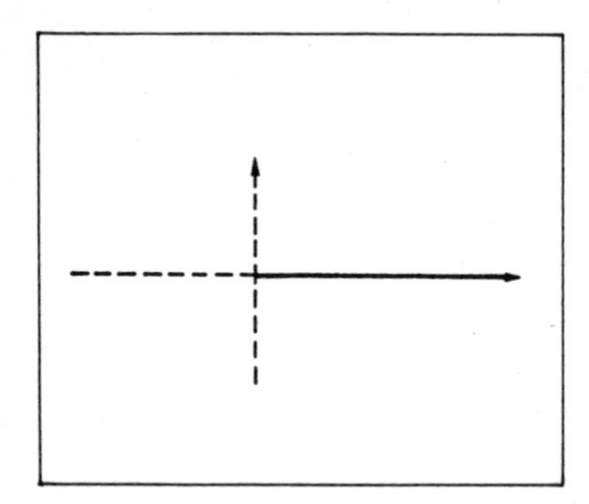

Fig. 2.

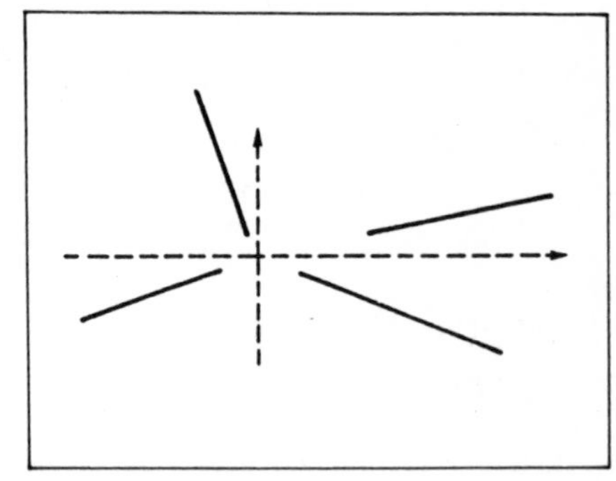

Fig. 3.

5. ISOPERIMETRIC INEQUALITY FOR p^*

Consider a cone like domain as in the previous section. In this case the critical number p^* depends on the least eigenvalue ω of the domain Ω on the sphere S^{N-1}. Let $\mathcal{O}(A)$ denote the class of all domains $\Omega \subseteq S^{N-1}$ of given $(N-1)$-dimensional Hausdorff measure A.

By means of Pólya and Szegö's *circular symmetrization* [PS] it follows that among all $\Omega \subset \mathcal{O}(A)$ the circle Ω^* yields the minimal value of ω.

Inserting this inequality into the expression for p^*

$$p^* = 1 + 2(2 + 2q + \sigma)/\{N + 2 + \sqrt{(N-2)^2 + 4\omega}\}$$

we obtain.

THEOREM 5.1. *Among all cone like domains $\mathcal{D}$ such that meas $\Omega = A$, the circular cone yealds the highest value of p^*.*

This result agrees with what is known for bounded domains, namely symmetrization decreases the stability.

6. STATIONARY STATES

We turn to the discussion of the stationary states of problem (P) in cone like domains in the case where $q = 0$. The equation for these stationary states in

$$\Delta u + r^\sigma u^p = 0 \text{ in } \mathcal{D}, u = 0 \text{ in } \partial\mathcal{D}, u > 0 \text{ in } \mathcal{D}. \tag{6.1}$$

The question wether or not (6.1) possesses positive solutions isn't yet completely solved.

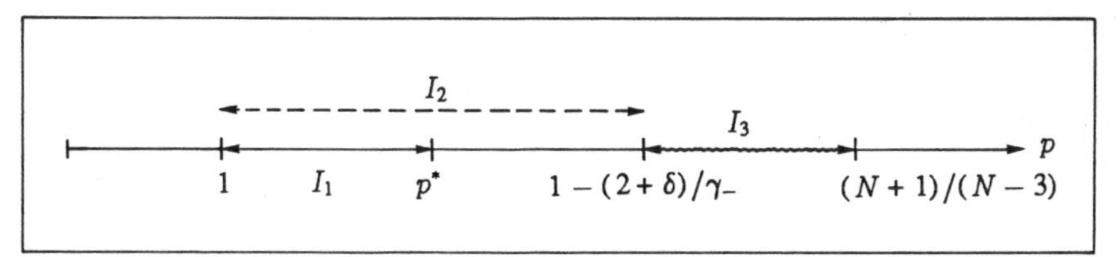

Fig. 4.

From theorem 4.1 it follows that no solutions exist for $p \in \left(1, 1 + \dfrac{2+\sigma}{2-\gamma_-}\right) =: I_1$.

Rather elementary considerations show [BL1, BE] that there are no stationary solutions for $p \in \left(1, 1 - \dfrac{2+\sigma}{\gamma_-}\right) =: I_2$.

By means of variational methods it is shown [BL1] that in the interval

$$I_3 := \left(1 - \frac{2+\sigma}{\gamma_-}, \begin{cases} \frac{(N+1)}{(N-3)} & \text{if } N > 3 \\ \infty & \text{if } N = 2, 3 \end{cases}\right)$$

there exist singular solutions of the form

$$u_s(r,\theta) = r^{-\frac{2+\sigma}{p-1}} \alpha(\theta). \tag{6.2}$$

A further nonexistence result is proved in [BL1] : if $\sigma = 0$ and $1 - 2/\gamma_- < p < \begin{cases} \frac{(N+2)}{(N-2)} & \text{if } N > 2 \\ \infty & \text{if } N = 2 \end{cases}$, then there are no regular, nontrivial solutions $u \leq u_s$. It should be mentioned that for $N > 2$ and $p > \dfrac{N+2+2\sigma}{N-2}$ such solutions exist in the whole space $\mathcal{D} = \mathbb{R}^N$ and in the exterior of spheres $\mathcal{D} = \{x : |x| > R\}$. A complete picture of positive radially symmetric solutions is given in [BM]. More details on solutions of (6.1) in cones are found in [BE].

7. ASYMPTOTIC BEHAVIOUR OF GLOBAL SOLUTIONS

The global solutions $u(x,t)$ of problem (P) can be divided into two classes:

(1) $u(x,t) \leq c$ in $\mathcal{D} \times \mathbb{R}^+$ for some $c > 0$

(2) $\exists \{t_k\}_{k=1}^{\infty}, t_k \to \infty$ as $k \to \infty$ such that $\lim_{k\to\infty} |u(\cdot, t_k)|_\infty = \infty$.

Global solutions have been studied by various authors [F1, F2, M1, M2, LM, K, B, BL2]. Except for Kavian who used the theory of potential well, all treated this problem by introducing suitable upper solutions $\bar{u}(x,t)$. Recall that $\bar{u} \geq 0$ is called an upper solution of $\mathcal{D}$, it if satisfies

$$\bar{u}_t - \Delta \bar{u} \geq t^q r^\sigma \bar{u}^p \text{ in } \mathcal{D} \times (0,T) \ \bar{u}(x,0) \geq u_0(x) \tag{7.1}$$

By adapting Fujita's candidate we find [B, BL2].

LEMMA 7.1. Assume $p > 1 + \dfrac{2+2q+\sigma}{N}$. Then for any $t_0 > 0$ there exists a positive number $\gamma > 0$, independent of $\mathcal{D}$, such that

$$\bar{u}(x,t) = (t+t_0)^{-\gamma} e^{-\frac{r^2}{4(t+t_0)}} \tag{7.2}$$

is an upper solution for problems with

$$u_0(x) \leq t_0^{-\gamma} e^{\frac{-r^2}{4t_0}}. \tag{7.3}$$

From this lemma it follows that all solutions $u(x,t;u_0)$ such that $u_0 \leq \bar{u}(x,0)$ are global, belong to the class (1) and tend to zero as $t \to \infty$.

Levine and Meier [LM] constructed an upper solution for cone like domains in the case where $p > 1 + \dfrac{2+2q+\sigma}{2-\gamma_-}$. Their construction is based on an argument in [M2]. This upper solution has also the properties that it decays to zero as t tends to infinity and that for some $c > 0, e^{cr^2}\bar{u}(x,t) \to 0$ as $|x| \to \infty$.

It can be shown [BL2] that solutions $u(x,t,u_0)$ with $u_0(x) \geq \epsilon$ for $|x| > R$ and any $\epsilon > 0$ cannot be global. Hence the zero solution cannot be stable in the supremum norm topology.

It would be interesting to construct solutions belonging to the second class.

8. CONCLUSION

We are now faced with the problem to understand the geometrical properties of $\mathcal{D}$, responsible for the Fujita phenomenon (property (F)). For example do the solution of (P) in the domain $\mathcal{D}$ consisting of increasing spheres connected by thin handles behave like those in a cone or in a cylinder or arbitrarily? It is clear that in a «thin» domain the effect of the boundary can prevent the solutions from blowing up, whereas in «thick» domains the influence of the heat source is more perceptible. In view of Meier's result (Lemma 2.3) every progress in this direction is useful for the theory of the linear heat equation in unbounded domains.

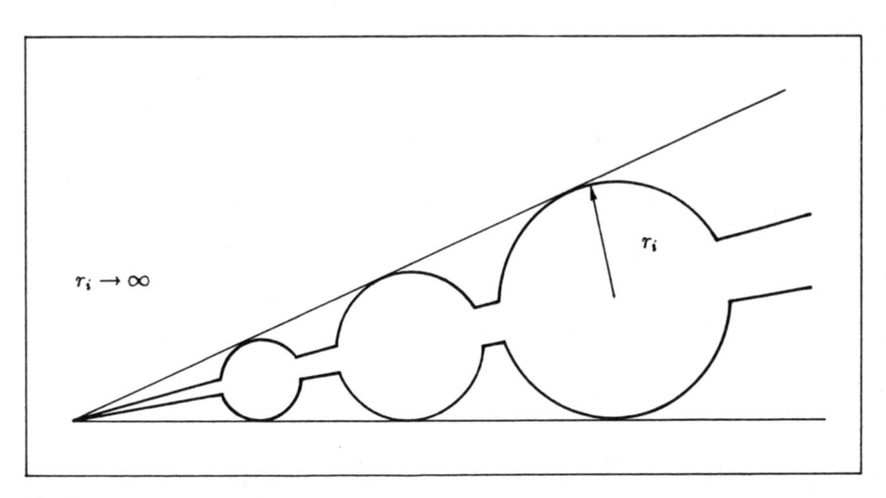

Fig. 5.

REFERENCES

[AW] D. J. ARONSON and H. F. WEINBERGER: Multidimensional nonlinear diffusion arising in population genetics, *Advanced in Math.* **38** (1978), 33-76.

[B] C. BANDLE: Blow up in exterior domains, «Proc. Nancy Conf. on Non-linear Evolution Equations» (1988).

[BE] C. BANDLE and M. ESSÈN: Positive solutions of Emden equations in cone-like domains.

[BL1] C. BANDLE and H. A. LEVINE: On the existence of global solutions of reaction-diffusion equations in sectorial domains, *Trans. Am. Math. Soc.* (to appear).

[BL2] C. BANDLE and H. A. LEVINE: Fujita type results for convective reaction-diffusion equations in exterior domains *ZAMP* **40** (1989), 665-676.

[BM] C. BANDLE and M. MARCUS: The positive radi al solutions of a class of semilinear elliptic equations *J. reine und angewandte Math* **401** (1989), 25-59.

[F1] H. FUJITA: On the blowing up of a solutions o the Cauchy problem for $u_t = \Delta u + u^{1+\alpha}$, *J. Fac. Sci., Tokyo Sec. IA, Math.* **13** (1966), 109-124.

[F2] H. FUJITA: On some nonexistence and nonuniqueness theorems for nonlinear parabolic equations, *Proc. Symp. Pure Math.* **18** (1969), 105-113.

[H] K. HAYAKAWA: On nonexistence of global solutions of some semilinear parabolic differential equations, *Proc. Japan Acad.* **49** (1973), 503-525.

[Kap] S. KAPLAN: On the growth of solutions of quasilinear parabolic equations, *Comm. Pure Appl. Math.* **16** (1963), 305-333.

[Kav] O. KAVIAN: Remarks on the large time behaviour of a nonlinear diffusion equation, *Ann. Institut Henri Poincaré*, «Analyse nonlinéaire» **4** (1987), 423-452.

[L1] H. A. LEVINE: The long time behaviour of solutions of reaction-diffusion equations in unbounded domains: a survey. «Proc. of the 10th Dundee Conference on the Theory of Ordinary and Partial Differential Equations» (1988).

[L2] H. A. LEVINE: Some nonexistence and instability theorems for solutions of formally parabolic equations of the form $Pu_t = Au + F(u)$, *Arch. Rat. Mech. Anal.* **51** (1973), 371-386.

[LM] H. A. LEVINE and P. MEIER: The value of the critical exponent for reaction-diffusion equations (to appear) in Arch. Rat. Mech. Math.

[M1] P. MEIER: Blow up of solutions of semilinear parabolic differential equations, *ZAMP* **39** (1988), 135-149.

[M2] P. MEIER: On the critical exponent for reaction-diffusion equations (to appear).

[PS] G. PÒLYA and SZEGÖ: Isoperimetric inequalities in mathematical physics, Princeton University Press (1951).

[W] F. WEISSLER: Existence and nonexistence of global solutions for a semilinear heat equation, *Israel J. Math.* **38** (1981), 29-40.

A FATOU THEOREM FOR THE p-LAPLACIAN

E. FABES - N. GAROFALO
S. MARIN-MALAVE - S. SALSA

1. INTRODUCTION

A classical Fatou theorem states:

For any given nonnegative harmonic function $u(x)$ defined in the unit ball D of $\mathbf{R}^n$ there exists a set of boundary points E_u, with surface measure equal to the surface measure of the entire boundary, such that for each point $P \in E_u$, $\lim_{\substack{x \to P \\ x \in B \cap \Gamma_P}} u(x)$ exists, where Γ_P is any finite cone with vertex P and interior contained in B. This behavior of u is often described by saying « u has a nontangential limit at almost every (with respect to surface measure) boundary point».

The above Fatou theorem has been generalized to nonnegative solutions of general second order elliptic equations with nonsmooth coefficients ([1], [2]). The statement of the Fatou theorem in these situations remains the same except that surface measure must be replaced by the harmonic measure associated with the governing partial differential operator.

In this work we wish to extend the Fatou theory to nonnegative solutions u of the p-Laplacian equation

$$\operatorname{div}(|\nabla u|^{p-2} \nabla u) = 0 \quad (1 < p < \infty)$$

Our Fatou theorem in this case takes the following form:

Let D be a unit ball in $\mathbf{R}^n$ and suppose u is a nonnegative weak solution of $\operatorname{div}(|\nabla|^{p-2} \nabla u) = 0$ in D. The set of boundary points E_u at which u has a nontangential limit has Hausdorf dimension $\geq \beta$, a positive number depending only on p and n. Recent examples of Tom Wolff ([10]) and John Lewis ([7]) show that the set E_u could have surface measure zero. The positive result at least guarantees that E_u is somewhat far from being empty.

The above Fatou theorem for the p-Laplacian in smooth domains of $\mathbf{R}^n$ and with $1 < p < 3 + \dfrac{2}{2-n}$ was first proved by Manfredi and Weitsman in [18]. Several ideas in their paper were very useful and we would like to thank the authors for sharing them with us early on in their work.

Our technique for establishing a Fatou theory in the nonlinear setting is itself linear in nature. If frequently occurs that results for linear equations if achieved in sufficient generality can be successfully applied to nonlinear situations. The present work is one more example of this phenomenum. In order to find a Fatou theorem in the nonlinear setting we first develop a potential theory for linear uniformly elliptic equations in nondivergence form and for the corresponding adjoint equations. The coefficients of the linear operator are assumed to be smooth, but, more importantly for applications, all the constants involved in the basic estimates do *not* depend in any quantitative manner on the smoothness of the coefficients. With regard to dependence on the coefficients, our constants will take into account only ellipticity and L^∞-bounds.

The primary results in the linear theory are what we refer to as Comparison Theorems for nonnegative solutions of either a nondivergence form elliptic equation or of associated adjoint equation. In brief terms these theorems state that two nonnegative solutions of the equation (or adjoint equation) which vanish on an open portion of the boundary must vanish there at the same rate. For a precise statement of these results see theorems (I. 1.8) and (I. 2. 1).

The paper is divided into two parts. Part I devoted to the potential theory for second order linear nonvariational elliptic operators and their adjoints, always aiming toward establishing the above mentioned Comparison Theorems, especially Theorem I. 1. 8. Part II applies the linear theory to prove the previously mentioned Fatou theorem for the p-Laplacian. Part II may be read somewhat independently of Part I, at least from the point of view of understanding how the linear theory enters in establishing the positive Hausdorf dimension of the set of boundary points at which the solution of the nonlinear problem has a nontangential limit.

PART I. POTENTIAL THEORY FOR NONDIVERGENCE FORM OPERATORS AND THEIR ADJOINTS

Before we begin the main body of this paper we would like to recall the basic definitions and introduce the primary notation which will be extensively used throughout the work.

DEFINITION A. A bounded domain D of $\mathbf{R}^n$ is called a Lipschitz domain if:
(i) for each $Q \in \partial D$ there exists a coordinate system $(x', x_n) \in \mathbf{R}^{n-1} \times \mathbf{R}$,

a number r_0 and a function $\varphi : \mathbf{R}^{n-1} \to \mathbb{R}$ satisfying $\|\nabla\varphi\|_{L^\infty(\mathbf{R}^{n-1})} \leq m$ such that

(ii) $B_{r_0}(Q) \cap D = \{(x', x_n : x_n > \varphi(x')\} \cap B_{r_0}(Q)$, $(B_{r_0}(Q)$ = the ball in $\mathbf{R}^n$ with center Q and radius $r_0)$ and

$$B_{r_0}(Q) \cap \partial D = \{(x', \varphi(x')\} \cap B_{r_0}(Q).$$

We may assume the numbers m and r_0 are the same for each $Q \in \partial D$ and we will say that these numbers determine the Lipschitz character of D.

Given the coordinate system about $Q \equiv (Q', Q_n) \in \partial D$, we define for $r \leq r_0$.

$$T_r(Q) \equiv \{(x', x_n) : |x' - Q'| < r, \ |x_n - Q_n| < mr\}$$
$$A_r \equiv T_r(Q) \cap \partial D, \psi_r(Q) = T_r(Q) \cap D,$$

and $A_r(Q) = (Q', Q_n + mr)$. When the point Q is understood and unimportant in the discussion we will use the notation T_r, Δ_r, ψ_r and A_r.

In Part I we consider elliptic operators of the form

$$Lu(x) = \sum_{i,j=1}^{n} a_{ij}(x) D^2_{x_1, x_j} u(x) \qquad (x \in \mathbb{R}^n).$$

We assume the coefficients are smooth and the matrix $a(x) \equiv (a_{ij}(x))$ is bounded, symmetric, and positive definite, uniformly in x, i.e. there exist positive number λ and Λ such that

$$\lambda|\xi|^2 \leq \sum_{i,j=1}^{n} a_{ij}(x)\xi_i\xi_j \leq \Lambda\,|\xi|^2$$

for all x and ξ in $\mathbf{R}^n$. We wish again to emphasize that the assuption of smoothness of $a_{ij}(x)$ is only a qualitative one. In our estimates the dependence of the constants on the coefficients will only be in terms of the ellipticity parameters λ and Λ and the dimension n.

Corresponding to the operator L we have the adjoint operator L^* defined by

$$L^* v(y) = \sum_{i,j=1}^{n} D^2_{y_1 y_j}(a_{ij}(y) v(y)).$$

For a given Lipschitz domain D we let $g_D(x, y) \equiv g_{D,L}(x, y)$ be the Green's function corresponding to the operator L and domain D. In particular $L(g_D(\cdot, y))$ $(x) = 0$ for $x \in D\backslash\{y\}$, $L^*(g_D(x, \cdot))(y) = 0$ for $y \in D\backslash\{x\}$, and $g(Q, y) = 0 = g(x, Q)$ for $Q \in \partial D, x \in D$, and $y \in D$.

Section I.1. The notion of a normalized adjoint solution and the boundary Harnack principle

DEFINITION B. For a ball $\mathcal{B}$ let $t\mathcal{B}$ denote the ball concentric with $\mathcal{B}$ and radius equal to t times the radius of $\mathcal{B}$. Assume $\frac{1}{4}\mathcal{B} \supset \bar{D}$ and fix a point $P \in \frac{3}{4}\mathcal{B}\setminus\frac{1}{2}\mathcal{B}$. A *normalized adjoint* solution for L^* *and* D (briefly n.a.s.) is any function $\tilde{v}$ of the form

$$\tilde{v}(y) = \frac{v(y)}{g_{\mathcal{B}}(P,y)}$$

where v is a solution of the adjoint equation $L^*v = 0$ in D. (We shall see that for our purposes there is no quantitative dependence of $\tilde{v}$ on $\mathcal{B}$ and $g_{\mathcal{B}}(P,y)$. Therefore we do not highlight them in the name.)

The notion of a normalized adjoint solution was used extensively in [1]. The main purpose of Part I is to refine the results there by dropping the dependence on the modulus of continuity of the coefficients. We begin by recalling some basic facts from [1] concerning these functions.

THEOREM (I. 1.1). (Interior Harnack principle for n.a.s.). Suppose $\tilde{v}$ is a non-negative n.a.s. in a ball B_{2R} of radius $2R$. There exists a constant C depending only on n, λ and Λ such that for all $s \le R$

$$\sup\{\tilde{v}(y); y \in B_s\} \le C \inf\{\tilde{v}(y); y \in B_s\}.$$

(B_s and B_{2R} are assumed concentric.)

THEOREM (I. 1. 2). (The Dirichlet problem for n.a.s.). Given $f \in C(\partial D)$ there exists a unique normalized adjoint solution, $\tilde{v}$, such that $\tilde{v} \in C(\bar{D})$ and $\tilde{v} = f$ on ∂D. Moreover the maximum and minimum values of $\tilde{v}$ occur on ∂D.

DEFINITION C. Under the conditions of Theorem I.1.2, for fixed $y \in D$, the map $f \to \tilde{v}(y)$ is a positive continuous linear functional on $C(\partial D)$. So by the Riesz Representation Theorem

$$v(y) = \int_{\partial D} f(Q)\tilde{\omega}^y(dQ)$$

where $\tilde{\omega}^y$ is a regular Borel measure on ∂D. We will call $\tilde{\omega}^y$ the normalized adjoint measure at y (corresponding to L^* and D).

Notice that when D is a smooth domain

$$\tilde{\omega}^y(dQ) = \frac{\partial}{\partial v_Q} g(Q,y) g_{\mathcal{B}}(P,Q) / g_{\mathcal{B}}(P,y) \cdot \sigma(dQ)$$

where g denotes the Green's function for L and D, v_Q is the inward conormal to ∂D at $Q \cdot (v_Q = a(Q)(N_Q)$ with $a = (a_{ij})$ and N_Q is the unit inward normal to ∂D at $Q)$, $\frac{\partial}{\partial v_Q} g(Q, y) = \nabla_x g(x, y)|_{x=Q} \cdot v_Q$, and $\sigma(\mathrm{d}Q)$ is surface measure. The normalized adjoint measure $\tilde{\omega}^y$ will play an important role in the study of normalized adjoint solutions.

LEMMA (I. 1.3). Let $\tilde{\omega}^y_{2r}$ denote the normalized adjoint measure corresponding to L^* and a ball B_{2r}. For $Q \in \partial B_{2r}$ and $0 < \delta < 1$ set $\Delta_{\delta r} = \partial B_{2r} \cap B_{\delta r}(Q)$. There exists a positive constant c depending only on the ellipticity parameters, n, and δ such that

$$\inf_{y \in B_r} \tilde{\omega}^y_{2r}(\Delta_{\delta r}) \geq c$$

(B_r is assumed concentric with B_{2r}.)

PROOF. By a translation we may assume B_{2r} is centered at the origin and by a dilation we assume $r = 1$.

$$\tilde{\omega}^y_2(\Delta_\delta) \int_{\Delta_\delta} \frac{\partial}{\partial v_Q} g_2(Q, y) g_B(P, Q) \sigma(\mathrm{d}Q) / g_B(P, y)$$

where $g_2(x, y)$ is the Green's function corresponding to B_2 and elliptic operator L. Fix a point $A \in \partial B_{3/2}$. All the succeding constants in this Lemma will depend at most on the dimension n, the parameters of ellipticity, λ and Λ, for L, and δ.

From Hopf's lemma and Harnack's inequality ([4], [5]) there exists $c_1 > 0$ such that for all $y \in B_1$, and $Q \in \partial B_2$,

$$\frac{\partial}{\partial v_Q} g_2(Q, y) \geq c_1 g_2(A, y).$$

(See [1, Lemma 4.3, p. 166]). Hence, for $y \in B_1$

$$\tilde{\omega}^y_2(\Delta_\delta) \geq c_1 g_2(A, y) \int_{\Delta_\delta} g_B(P, Q) \sigma(\mathrm{d}Q) / g_B(P, y).$$

Let $g_3(x, y)$ denote the Green's function for B_3 and L. Then Harnack's property for normalized solutions (Theorem (I. 1.1)) implies

$$\frac{g_B(P, Q)}{g_3(A, Q)} \geq c_2 \frac{g_B(P, y)}{g_3(A, y)}$$

for all $Q \in \partial B_2$ and $y \in B_1$. Hence

$$\tilde{\omega}^y_2(\Delta_\delta) \geq c_1 c_2 \frac{g_2(A, y)}{g_3(A, y)} \int_{\Delta_\delta} g_3(A, Q) \sigma(\mathrm{d}Q).$$

Once again using Harnack's inequality (for n.a.s.) we have for $y \in B_1$.

$$\frac{g_2(A,y)}{g_3(A,y)} \geq c_3 \frac{\int_{B_1} g_2(A,z)\,\mathrm{d}z}{\int_{B_1} g_3(A,z)\,\mathrm{d}z} \geq c_4 > 0.$$

We are now left to show:

LEMMA (I. 1.4). Let $g_3(x,y)$ denote the Green's function for L and B_3. For $Q \in \partial B_2$ and $0 < \delta < 1$, set $\alpha_\delta = \partial B_2 \cap B_\delta(Q)$. There exists a positive constant c depending only on n, λ, Λ, and δ such that for all $A \in \partial B_{3/2}$

$$\int_{\alpha_\delta} g_3(A,Q)\sigma(\mathrm{d}Q) \geq c > 0.$$

PROOF. Let $D \subset B_2 \backslash B_{7/4}$ be a smooth such that $\alpha_\delta \subset \partial D$ and D contains a ball B' whose radius depends only on δ. Now let $D' \subset B_3$ be another smooth domain containing D such that $\alpha_\delta \subset D'$ and $\partial D' \cap \partial D = \partial D \backslash \alpha_\delta$. We pick a point $A' \in D' \backslash \bar{D}$ so that the positive constants in the following chain of inequalities depend only on the parameters of ellipticity, λ and Λ, the dimension n, and δ :

$$\int_{\alpha_\delta} g_3(A,Q)\sigma(\mathrm{d}Q) \geq c_1 \int_{\alpha_\delta} g_3(A',Q)\sigma(\mathrm{d}Q) \geq c_2 \int_{\alpha_\delta} g_D(A',Q)\sigma(\mathrm{d}Q).$$

In the final inequality above g_D, denotes the Green's function for D and L and the inequality results from the fact $g_{D'} \leq g_3$ since $D' \subset B_3$.

Since $\partial D \backslash \alpha_\delta = \partial D \cap \partial D'$, $g_{D'}(A',Q) = 0$ for $Q \in \partial D \backslash \alpha_\delta$. Therefore

$$\int_{\alpha_\delta} g_{D'}(A',Q)\sigma(\mathrm{d}Q) = \int_{\alpha_D} g_{D'}(A',Q)\sigma(\mathrm{d}Q).$$

The function $g_{D'}(A',y)$, as a function of y, satisfies $L^*(g_{D'}(A',\cdot)(y) = 0$ for $y \in D$. The function $W(x) = -\int_{B'} g_D(x,y)\,\mathrm{d}y$ satisfies $LW = \chi_{B'}$, the characteristic function of B', and $W|_{\partial D} = 0$. Hence, an integration by parts gives

$$\int_{B'} g_{D'}(A',y)\,\mathrm{d}y = \int_D g_{D'}(A',y)LW(y)\,\mathrm{d}y =$$
$$= \int_{\partial D} g_{D'}(A',Q)\frac{\partial}{\partial v_Q}W(Q)\sigma(\mathrm{d}Q).$$

Again from [1, Lemma 4.3, p. 166] there exists $C > 0$ such that

$$\frac{\partial W}{\partial v_Q}(Q) \leq C \sup_D \int_{B'} g_D(x,y)\,\mathrm{d}\,y \leq \tilde{C}$$

$\tilde{C}$ depending only on λ, Λ, n, and δ. We now have

$$\int_{\alpha_\delta} g_{D'}(A', Q)\sigma(\mathrm{d}\,Q) \geq c \int_{B'} g_{D'}(A', y)\,\mathrm{d}\,y,$$

and using a maximum principle argument the last integral is bounded below by a positive constant depending only λ, Λ, n, and δ. (See [4, Proof of Lemma 3.3].) This concludes the proof of Lemma I.1.4 and so also the proof of Lemma I.1.3.

As a consequence of Lemma I.1.3 we obtain the Hölder continuity of a nonnegative normalized adjoint solution at the open parts of the boundary where it vanishes.

THEOREM (I. 1.5). Let D be a Lipschitz domain in $\mathbb{R}^n$ with constants r_0 and m determining the Lipschitz character of D. Let $L = \sum_{i,j=1}^n a_{ij}(x) D^2_{x_i x_j}$ be a uniformly elliptic operator with parameters of ellipticity λ and Λ. Fix $Q \in \partial D$, $r \leq r_0/2$, and assume $\tilde{v}$ is a nonnegative normalized adjoint solution for L^* in $\psi_{2r} \equiv \psi_{2r}(Q)$ which continuosly vanishes on $\Delta_{2r} \equiv \Delta_{2r}(Q)$. There exists a $\theta, 0 < \theta < 1$ depending only on λ, Λ, n, and m such that

$$\sup_{\psi} \tilde{v} \leq \theta \sup_{\psi_{2r}} \tilde{v}.$$

As a consequence there exist positive constants C and α depending only on the above parameters such that for all $y \in \psi_{2r}$,

$$\tilde{v}(y) \leq c \left(\frac{|y-Q|}{r}\right)^{\alpha} \sup_{\psi_{2r}} \tilde{v}.$$

PROOF. We may assume $\sup_{\psi_{2r}} \tilde{v} = 1$. Let $\tilde{\omega}^y_{2r}$ denote the normalized adjoint measure for L^* and $B_{2r} \equiv B_{2r}(Q)$. We can find a positive number δ, depending on m, and a point $\tilde{Q} \in \partial B_{2r}$ such that

$$\alpha_{\delta r} \equiv B_{\delta r}(\tilde{Q}) \cap \partial B_{2r} \subset \partial B_{2r} \backslash \bar{D}.$$

Form the maximum principle for n.a.s. (Theorem (I.1.2))

$$1 - \tilde{v}(y) \geq \tilde{\omega}^{y}_{2r}(\alpha_{\delta r}).$$

Therefore, using Lemma I. 1. 4.

$$\inf_{\psi_r} (1 - \tilde{v}(y)) \geq c > 0$$

with c depending only on λ, Λ, n, and m. Then

$$\sup_{\psi_r} v \leq 1 - c \equiv \theta.$$

To conclude the Hölder continuity we observe that the last inequality above implies: for all $y \in \psi_{2^{-k}r}(Q)$,

$$\tilde{v}(y) \leq \theta^{k+1} \sup_{\psi_{2r}}, \; k = 0, 1, 2, \ldots.$$

This immediately gives the existence of positive numbers c and α depending only on λ, Λ, n, m, such that for $y \in \psi_{2r}(Q)$

$$\tilde{v}(y) \leq c \left(\frac{|y - Q|}{r} \right)^{\alpha} \sup_{\psi_{2r}} \tilde{v}.$$

Theorems (I. 1.1) and (I. 1.5) allow us to repart verbatim the arguments given in [1, Lemma 2.4, p. 157] to prove a strengthened version of (I.1.5).

THEOREM (I. 1.6). (Boundary Harnack principle for n.a.s.). Under the hypotheses of Theorem I. 1.5, assume again $\tilde{v}$ is a nonnegative normalized adjoint solution for L^* and $\psi_{2r}(Q)$ which vanishes continuously on $\Delta_{2r}(Q), Q \in \partial D$. There exist positive constant C and α depending only on λ, Λ, n, and m, such that for $y \in \psi_r(Q)$

$$\tilde{v}(y) \leq C \left(\frac{|y - Q|}{r} \right)^{\alpha} \tilde{v}(A_r(Q))$$

where, recall for $Q = (Q', Q_n), A_r(Q) = (Q', Q_n + mr)$.

DEFINITION D. For the remainder of the paper we will assume D is a bounded C^2-domain of R^n. This means there exists a positive number m such that in the local representation of ∂D as a graph $(x', \varphi(x')), x' \in R^{n-1}$, we have $\varphi \in C^2(R^{n-1})$ and

$$\sum_{1 \le |\alpha| \le 2} \|D^\alpha \varphi\|_{L^\infty(R^{n-1})} \le m. \quad \text{(See definition A.)}$$

From now on we will refer to this number m and the number r_0 in Definition A as determining the smoothness or C^2 character of D.

In what follows we will also say that two objects A and B (numbers of functions) are equivalent and write $A \sim B$ if there exists a positive constant C depending at most on ellipticity parameters, dimension, and the smoothness character of D such that

$$\frac{1}{C} A \le B \le CA.$$

THEOREM (I. 1.7). (Comparison theorem for n.a.s). Let D be a bounded C^2-domain of R^n. Assume L is a uniformly elliptic operator with parameters of ellipticity λ and Λ (and having smooth coefficients). Let $\tilde{v}$ and $\tilde{w}$ be two nonnegative normalized adjoint solutions in $\psi_{2r}(Q), Q \in \partial D$, continuously vanishing on $\Delta_{2r}(Q)$ $(r \le r_0 \backslash 2)$. Then for $y \in \psi_r(Q)$,

$$\frac{\tilde{v}(y)}{\tilde{w}(y)} \sim \frac{\tilde{v}(A_r(Q))}{\tilde{w}(A_r(Q))}.$$

PROOF. We construct a C^2-domain D_r satisfying: $\psi_{5r/4}(Q) \subset D_r \subset \psi_{2r}(Q)$, $\partial D_r \cap \partial D \supset \Delta_{3r/2}(Q)$, and the C^2 character of D_r is controlled by the C^2 character of D (and therefore independent of r). Let $G_r(x, y)$ denote the Green's function corresponding to L and D_r.

For $y \in \psi_r(Q)$,

$$\begin{aligned} \tilde{v}(y) &= \int_{\partial D_r} \frac{\partial}{\partial v_Q} G_r(Q, y) \tilde{v}(Q) \mathrm{d}\sigma(Q) / G_B(P, y) \\ &\sim \frac{1}{r} \frac{G_r(A_{5r/2}(Q), y)}{G_B(P, y)} \int_{\partial D_r} \tilde{v}(Q) \mathrm{d}\sigma(Q). \end{aligned}$$

(See [1, Lemma 4.3. p. 166].) Since the same equivalence-type relation holds for $\tilde{w}$, we have for $y \in \psi_r(Q)$,

$$\frac{\tilde{v}(y)}{\tilde{w}(y)} \sim \int_{\partial D_r} \tilde{v}(Q) \mathrm{d}\sigma(Q) / \int_{\partial D_r} \tilde{w}(Q) \mathrm{d}\sigma(Q).$$

The right-hand side of the above equivalence does not depend on $y \in \psi_r(Q)$ and so any two values of $\frac{\tilde{v}}{\tilde{w}}$ are equivalent; in particular for $y \in \psi_r(Q)$,

$$\frac{\tilde{v}(y)}{\tilde{w}(y)} \sim \frac{\tilde{v}(Ar(Q))}{\tilde{w}(Ar(Q))}.$$

THEOREM (I. 1.8). (Comparison theorem for adjoint solutions). Let D and L satisfy the hypotheses of Theorem (I.1.7). Let v and w be two nonnegative adjoint solutions, i.e. $L^*v = L^*w = 0$, in $\psi_{2r}(Q), Q \in \partial D$, continuously vanishing on $\Delta_{2r}(Q)(r \leq r_0/2)$. Then for $y \in \psi_r(Q)$,

$$\frac{v(y)}{w(y)} \sim \frac{v(Ar(Q))}{w(Ar(Q))}.$$

PROOF. The functions $\tilde{v} \equiv \dfrac{v(y)}{G_B(P,y)}$ and $\tilde{w}(y) \equiv \dfrac{w(y)}{G_B(P,y)}$ are normalized adjoint solutions satisfying the hypotheses of Theorem I.1.7. Hence the conclusion of I. 1.8 holds for $\dfrac{\tilde{v}(y)}{\tilde{w}(y)}$. But $\dfrac{\tilde{v}}{\tilde{w}} \equiv \dfrac{v}{w}$.

COROLLARY (I. 1.9). Let v and w satisfy the hypothesis of Theorem I.1.8 in $\psi_{r_1}(Q)$ with $Q \in \partial D$ and $r_1 \leq r_0$. Then there exist positive constants C and α depending on λ, Λ, n, and m such that

$$\left| \frac{v(y)}{w(y)} - \frac{v(y')}{w(y')} \right| \leq C \frac{v(A_{r_1}(Q))}{w(A_{r_1}(Q))} \left(\frac{|y-y'|}{r_1} \right)^{\alpha}$$

for all y and $y' \in \psi_{r_1/2}(Q) \cap D$.

PROOF. The proof follows the method of Moser who obtains Hölder continuity from the uniform Harnack principle ([9]). In fact that argument and Theorem (I.1.5) give the above result when $y \in D \cap \overline{\psi_{r_1/2}(Q)}$ and $y' \in B_{s/2}(y)$, with $s = \operatorname{dist}(y, \partial D)$. Hence it is enough to show the Hölder continuity at a boundary point $Q_0 \in A_{r_1/2}/(Q)$. We may also assume $v(A_{r_1}(Q)) = w(A_{r_1}(Q)) = 1$.

Set

$$M(s) = \sup \left\{ \frac{v(y)}{w(y)};\ y \in \psi_s(Q_0) \right\}$$

and

$$m(s) = \inf \left\{ \frac{v(y)}{w(y)};\ y \in \psi_s(Q_0) \right\}.$$

Then for $s \leq r_1/4$,

$$M(s) - \frac{v}{w} = \frac{M(s)w - v}{w} \quad \text{and} \quad \frac{v}{w} - m(s) = \frac{v - m(s)w}{w}$$

are quotients of positive adjoint solutions in $\psi_s(Q_0)$ which vanish on $\Delta_s(Q_0)$. By theorem I. 1.8,

$$\sup_{\psi_{s/2}(Q_0)} (M(s) - \frac{v}{w}) \leq C \inf_{\psi_{s/2}(Q_0)} (M(s) - \frac{v}{w})$$

and

$$\sup_{\psi_{s/2}(Q_0)} (\frac{v}{w} - m(s)) \leq C \inf_{\psi_{s/2}(Q_0)} (\frac{v}{w} - m(s))$$

i.e.

$$M(s) - m(s/2) \leq C(M(s) - M(s/2))$$

and

$$M(s/2) - m(s) \leq C(m(s/2) - m(s)).$$

Adding the final two inequalities we obtain

$$M(s/2) - m(s/2) \leq \theta(M(s) - m(s))$$

where $0 < \theta \equiv \frac{C-1}{C+1} < 1$. Integrating gives

$$M(s) - m(s) \leq \theta^{-1} \left(\frac{s}{r_1}\right)^{\alpha} (M(r_1) - m(r_1))$$

with $\alpha = -\log_2 \theta$. This concludes the proof of the Corollary.

When v and w represent the Green's function for an elliptic operator at two fixed poles we obtain the following important

THEOREM (I. 1.10). Assume $L \equiv \sum a_{ij}(x) D^2_{x_i x_j}$, an elliptic operator, and D a bounded C^2 domain, satisfy the hypotheses of Theorem I. 1.7. Let $g(x, y)$ denote the Green's function corresponding to L and D. If x and x_0 are fixed points of D there exist positive C and α, α depending only on $\lambda, \Lambda, n, m, r_0$ but C depending in addition on $\delta = \min\{\text{dist}(x, \partial), \text{dist}(x_0, \partial D)\}$, such that

$$\left| \frac{g(x,y)}{g(x_0,y)} - \frac{g(x,y')}{g(x_0,y')} \right| \leq C|y' - y|^{\alpha}$$

for all y and y' belonging to $\{z \in D : \text{dist}(z, \partial D) \leq \delta/2\}$.

Section 2. Potential Theory for solution of $Lu = 0$; the doubling property for L-harmonic measure

We remind the reader that $L \equiv \sum_{i,j=1}^{n} a_{ij}(x) D^2_{x_i x_j}$ is an elliptic operator satisfying

$$\lambda |\xi|^2 \leq \sum_{i,j} (x) \xi_i \xi_j \leq \Lambda |\xi|^2$$

for all x and ξ in R^n (λ and Λ are positive constants). Also D in a C^2-domain with m and r_0 describing the smoothness character. (See Definitions A and D.)

THEOREM (I. 2.1). (Comparison theorem for solutions of $Lu = 0$). Let $Q \in \partial D$ and assume u_1 and u_2 are positive solutions of $Lu = 0$ in $\psi_{2r}(Q)$ which vanish on $\Delta_{2r}(Q), r \leq r_0$. Then for $x \in \psi_r(Q)$,

$$\frac{u_1(x)}{u_2(x)} \sim \frac{u_1(A_r(Q))}{u_2(A_r(Q))}.$$

PROOF. From Hopf's Lemma and Harnack's inequality, for $x \in \psi_r(Q)$,

$$u_1(x) \sim u_1(A_r(Q)) \quad \frac{\text{dist}(x, \partial D)}{r}.$$

(See the proof of Lemma 2.5 p. 158 in [1].) Hence I. 2.1 follows.

DEFINITION E. Given an elliptic operator $L \equiv \sum_{i,j} a_{i,j}(x) D^2_{x_i x_j}$ with smooth coefficients and a bounded Lipschitz domain D, for any given $g \in C(\partial D)$ there exists a unique solution u to the Dirichlet problem

$$Lu = 0 \text{ in } D, \ u|_{\partial D} = g.$$

Fixing a point $x \in D$, the maximum principle implies the map $g \to u(x)$ is a positive linear functional on $C(\partial D)$. Hence there exist a unique regular Borel measure ω^x on ∂D such that every $g \in C(\partial D)$

$$u(x) = \int_{\partial D} g(Q) \omega^x(dQ).$$

The measure $\omega^x(\partial D)$ is called the L-harmonic measure for D evaluated at x. At times we will emphasize the dependence of ω^x on L and D by using the notation $\omega^x_{L,D}$ or ω^x_D.

A consequence of the Comparison Theorem (I. 2.1) for solutions is

THEOREM (I. 2.2). (Doubling property for L-harmonic measure). Let ω^x denote the L-harmonic measure for a bounded C^2-domain D evaluated at x. Take $r \le r_0/4$ (r_0 as in Definition A). Then for all $Q \in \partial D$ and $D\backslash\psi_{2r}(Q)$

$$\omega^x(\Delta_r(Q)) \sim \omega^x(\Delta_{2r}(Q)).$$

PROOF. By applying Theorem (I.2.1) and the maximum principle to the functions $\omega^x(\Delta_{2r})$ and $\omega^x(\Delta_r)$ is the domain $D\backslash\psi_{4r}$, we obtain

$$\omega^x(\Delta_{2r}) \le C\,\frac{\omega^{A_{4r}}(\Delta_{2r})}{\omega^{A_{4r}}(\Delta_r)}\,\omega^x(\Delta_r)$$

for all $x \in D\backslash\Psi_{4r}$. There exist positive constants C and α, depending on λ, Λ, n, and m,such that for $x \in \psi_{r/2}$,

$$0 \le 1 - \omega^x(\Delta_r) \le C\left(\frac{|x-Q|}{r}\right)$$

and

$$0 \le 1 - \omega^x(\Delta_{2r}) \le C\left(\frac{|x-Q|}{r}\right).$$

This observation and Harnack's inequality ([4], [5]) imply $\omega^{A_{4r}}(\Delta_{2r})$ and $\omega^{A_{4r}}(\Delta_r)$ are each equivalent to an absolute constant depending on λ, Λ, n, and m.

Other consequences of Theorem (I.2.2) are concerned with the so-called kernel function. Proofs can be found in [2].

DEFINITION F. The kernel function $K(x, Q)$, normalized at $x_0 \in D$, is defined for $x \in D$ and $Q \in \partial D$ by

$$K(x,Q) = \lim_{y \to Q} \frac{g_D(x,y)}{g_D(x_0,y)}$$

where g_D denotes the Green's function corresponding to L and D. This is the same as

$$K(x,Q) = \frac{d\omega^x}{d\omega^{x_0}}(Q).$$

In fact we have the following characterization of the kernel function:

THEOREM (I. 2. 3). The kernel function is uniquely determined by the conditions:

a) $L(K(\cdot, Q))(x) = 0$ for $x \in D$,

b) $K(x_0, Q) = 1$ for all $Q \in \partial D$,

c) if $Q' \in \partial D$ and $Q' \neq Q$, then $\lim_{x \to Q'} K(x, Q) = 0$.

The relationships between L-harmonic measure evaluated at x_0 and the kernel function are given in the next theorem.

THEOREM (I. 2. 4). Let $K(x, Q)$ be the kernel function for L and D, normalized at x_0, and set $\omega^x \equiv \omega^x_{L,D}$. Then there exists a sequence of constants $\{C_j\}$ such that C_j depends only on $\lambda, \Lambda, n, m, r_0$, and $\operatorname{dist}(x_0, \partial D)$, $\sum C_j < \infty$, and

(i) $K(A_r(Q), Q) \sim \omega^{x_0}(\Delta_r(Q))^{-1}$ $(Q \in \partial D, r \leq \frac{r_0}{2})$

(ii) for t sufficiently small and all $\tilde{Q} \in \partial D$

$$\sup\{K(A_t(\tilde{Q}), Q) \in \Delta_{2^{j+1}t}(\tilde{Q}) \backslash \Delta_{2^j t}(\tilde{Q})\} \leq \frac{C_j}{\omega^{x_0}(\Delta_{2^j t}(\tilde{Q}))}.$$

We conclude Section I.2 with the statement of the Hölder continuity of $K(x, \cdot)$ on ∂D with x fixed in D. The Hölder continuity is an immediate corollary of Theorem I.1.10 and it together with the properties of K described in Theorem I. 2.4 form the main ingredients of the proof of the Fatou theorem for the L-Laplacian.

THEOREM (I. 2. 5). Assume $L \equiv \sum_{i,j=1}^{n} a_{i,j}(x) D^2_{x_i x_j}$ is an elliptic operator with smooth coefficients and parameters of ellipticity λ and Λ. Assume also D is bounded C^2-domain in R^n with smoothness characteristics m and r_0. Then there exist positive constants C and α, α depending only on $\lambda, \Lambda, n, m, r_0$ but C depending in addition on $\delta \equiv \min\{\operatorname{dist}(x, \partial D), \operatorname{dist}(x_0, \partial D)\}$ such that

$$|K(x, Q) - K(x, Q')| \leq C|Q - Q'|^{\alpha}$$

for all Q and Q' on ∂D. (Here K is normalized at x_0.)

PART II. A FATOU THEOREM FOR THE p-LAPLACIAN.

In Part II, we take D to be the unit ball in R^n with center the origin.

DEFINITION G. Fix $1 < p < \infty$. A function $u(x)$, $x \in D$, is a solution of the equation

$$\text{(II.1)} \qquad \Delta_p u \equiv \operatorname{div}(|\nabla u|^{p-2} \nabla u) = 0 \quad \text{in} \quad D$$

if $u \in W^{1,p}_{\text{loc}}(D) \equiv \{u \in L^p_{\text{loc}}(D) : \nabla u \in L^p_{\text{loc}}(D)\}$ and

$$\int_D |\nabla u|^{p-2}\nabla u \cdot \nabla\varphi \, d\,x = 0$$

for all $\varphi \in C_0^\infty(D)$. A function u which is a solution of II.1 is called a p-harmonic function and Δ_p is called the p-Laplacian.

Despite the degenerate character of II. 1 solutions belong to $C^{1,\alpha}_{\text{loc}}(D)$ with $\alpha = \alpha(p, n) > 0$. (See [3] and [6]). Concerning boundary behavior we have

THEOREM (II. 2). Let u be a nonnegative p-harmonic function in the unit ball D. Then $\{Q \in \partial D : u$ has a nontangential limit at $Q\}$ has Hausdorf dimension $\geq \beta > 0$ where β depends only on p and n.

PROOF. For $0 < r < 1$, we set $u_r(x) = u(rx)$ and obtain a p-harmonic function u_r in D which is $C^{1,\alpha}$ in $\bar{D}$, the closure of D. Now consider the reguralized p-harmonic operator

$$\text{(II.3)} \qquad \Delta_p^\epsilon v \equiv \operatorname{div}((|\nabla v|^2 + \epsilon)^{\frac{p-2}{2}}\nabla v), \quad \epsilon > 0;$$

and denote by $v_{\epsilon,r}$ the unique solution of the Dirichlet problem

$$\Delta_p^\epsilon v = 0 \text{ in } D, v|_{\partial D} = u_r|_{\partial D}.$$

Because of the nondegeneracy of Δ_p^ϵ, $v_{\epsilon,r}$ belongs to $C^\infty(D)$. Therefore we can perform the differentiation in II. 3 and, after dividing by $(\epsilon + |\nabla v_{\epsilon,r}|^2)^{\frac{p-2}{2}}$, we see that $v_{\epsilon,r}$ is a solution of the linear equation

$$L^u_{\epsilon,r}w \equiv \sum_{i,j=1}^n \left(\delta_{i,j} + (p-2)\,\frac{D_{x_i}v_{\epsilon,r}D_{x_j}v_{\epsilon,r}}{\epsilon + |\nabla v_{\epsilon,r}|^2}\right) D^2_{x_i x_j} w = 0\,.$$

(We have added the superscript u to the operator to keep in mind the dependence of the coefficients on the p-harmonic function u.) The operator $L^u_{\epsilon,r}$ has parameters of ellipticity λ, Λ depending only on p; namely, $\lambda = \min(1, p-1)$ and $\Lambda = \min(1, p-1)$, and the coefficients belong to $C^\infty(D)$.

A key property of $v_{\epsilon,r}$ is that $v_{\epsilon,r} \to u_r$ in $W^{1,p}(D)$ as $\epsilon \to 0$ ([6]). Also as $\epsilon \to 0$ $v_{\epsilon,r} \to u_r$ unformly on compact subsets of D.

We make one more dilation: the function $v_{\epsilon,r}(rx)$ satisfies an elliptic equation $\tilde{L}^u_{\epsilon,r}w = 0$ of the same form as that defined by $L^u_{\epsilon,r}$. The parameters of ellipticily can be taken dependent only on p and the coefficients now belong to $C^\infty(\bar{D})$. For $v_{\epsilon,r}(r,x)$ we have the following representation formula:

$$v_{\epsilon,r}(rx) = \int_{\partial D} v_{\epsilon,r}(rQ)\tilde{K}^u_{\epsilon,r}(x,Q)\tilde{\omega}^u_{\epsilon,r}(dQ) \tag{II.4}$$

where $\tilde{\omega}^u_{\epsilon,r}$ denotes the $\tilde{L}^u_{\epsilon,r}$-harmonic measure for D evaluated at the origin and $\tilde{K}^u_{\epsilon,r}(x,Q)$ denotes the kernel function associated with $\tilde{L}^u_{\epsilon,r}$ and D, normalized at the origin.

Since $\int_{\partial D}\tilde{\omega}^u_{\epsilon,r}(dQ) = 1$, we can select a sequence $\epsilon_j \to 0$ such that $\tilde{\omega}^u_{\epsilon_j,r}$ (dQ) converges weakly (as $j \to \infty$) to a regular Borel measure $\tilde{\omega}^u_r(dQ)$. Also, from the local uniform convergence of $v_{\epsilon,r}$ to u_r in D, $\lim_{j\to\infty} v_{\epsilon_j,r}(rQ) = u(r^2Q)$ uniformly on ∂D. Finally recall that each kernel function $\tilde{K}^u_{\epsilon,r}(x,Q)$ is a solution in x of a second order elliptic equation with parameters of ellipticity depending on p. Since $\tilde{K}^u_{\epsilon,r}(O,Q) = 1$, $\tilde{K}^u_{\epsilon,r}(x,Q)$ is locally Hölder continuous in x with local Hölder exponent and norm independent of Q, ϵ, r, and u, Also from Theorem I. 2. 5 when x varies over a fixed compact subset of D, $\tilde{K}^u_{\epsilon,r}(x,Q)$ is Hölder continuous on ∂D, with Hölder exponent depending only on p and n, and Hölder norm bounded independently of ϵ and r. Hence we may assume the sequence $\{K^u_{\epsilon_j,r}(x,Q)\}$ converges uniformly on ∂D for each fixed $x \in D$. The function

$$\tilde{K}^u_r(x,Q) = \lim_{j\to\infty} \tilde{K}^u_{\epsilon_j,r}(x,Q)$$

satisfies the same Hölder continuity just described for each kernel function on the sequence. We can now allow ϵ_j to tend to zero in II. 1. 4 and obtain the representation

$$u(r^2x) = \int_{\partial D} u(r^2Q)\tilde{K}^u_r(x,Q)\tilde{\omega}^u_r(dQ). \tag{II.5}$$

We repeat once more the arguments of the previous paragraph: $u(0) = \int_{\partial D} u(r^2Q)\tilde{\omega}^u_r(dQ)$ and $1 = \int_{\partial D}\tilde{\omega}^u_r(dQ)$ imply there exist a sequence $r_j \nearrow 1$ and regular Borel measures $\mu^u(dQ)$ and $\tilde{\omega}^u(dQ)$ such that as $r_j \nearrow 1$, $u(r_j^2Q)\tilde{\omega}^u_{r_j}$ (dQ) converges weakly to $\mu^u(dQ)$ while $\tilde{\omega}^u_{r_j}(dQ)$ converges weakly to $\tilde{\omega}^u$ (dQ). As already noted in the previous paragraph $\tilde{K}^u_r(x,Q)$ is Hölder continuous in x and Q for $Q \in \partial D$ and x restricted to a compact subset of D. The Hölder norm can be bounded and the Hölder exponent can be written independently of r. Hence as $r_j \nearrow 1$ we may assume

$$\tilde{K}^u_{r_j}(x,Q) \to \tilde{K}^u(x,Q)$$

uniformly on ∂D for each $x \in D$. Letting $r_j \nearrow 1$ we obtain the final representation

$$u(x) = \int_{\partial D} \tilde{K}^u(x, Q)\mu^u(\mathrm{d}Q). \tag{II.6}$$

The measure $\tilde{\omega}^u_{\epsilon,r}(\mathrm{d}Q)$ enjoy the doubling condition of Theorem (I. 2.2.) (for $x = 0$) with a doubling constant dependent only on p and n.The same property holds for the weak limits $\tilde{\omega}^u_r$ and $\tilde{\omega}^u$. On the other hand, the relationship between $\tilde{K}^u_{\epsilon,r}$ and $\tilde{\omega}^u_{\epsilon,r}$ expressed in Theorem (I. 2. 5) carry over to $\tilde{K}^u$ and $\tilde{\omega}^u$.

These relationship and the doubling property of $\tilde{\omega}^u$ allow us to apply the procedure in [2] obtaining the existence of non-tangential limits of u at all points $Q \in \partial D$ except for a set of $\tilde{\omega}^u$ measure zero.

In particular $\tilde{\omega}^u(\{Q \in \partial D : u \text{ has nontangential limit at } Q\}) = \tilde{\omega}^u(\partial D) = 1$. But the doubling property of $\tilde{\omega}_u$; namely,

$$\tilde{\omega}^u(\Delta_{2r}) \leq C\tilde{\omega}^u(\Delta_r) \quad (r \leq r_0)$$

with C depending only on p and n implies there exists positive constants C and β also depending only on these parameters such that

$$\tilde{\omega}^u(\Delta_r) \leq Cr^\beta \quad (r \leq r_0).$$

In particular the Hausdorf dimension of the set $\{Q \in \partial D : u$ has nontangential limit at $Q\}$ is $\geq \beta$.

REFERENCES

[1] P. Bauman: Positive solutions of elliptic equations in nondivergence form and their adjoints, *Arkiv for Matematik*, **22** (1984), 153-173.

[2] L. Caffarelli, F. Fabes, S. Mortola and S. Salsa: Boundary behavior of non-negative solutions of elliptic operators in divergence form, *Indiana J. Math.*, **30** (1981), 621-640.

[3] L. C. Evans: A new proof of local $C^{1,\alpha}$ regularity for solutions of certain degenerate elliptic P.D.E., *J. Differential Equations*, **45** (1982), 356-373.

[4] E. B. Fabes and D. W. Stroock: The L^p-integrability of Green's functions and fundamental solutions for elliptic and parabolic equations, *Duke Math. J.*, **51** (1984), 997-1016.

[5] N. V. Krylov and M. V. Safanov: A certain property of solutions of parabolic equations with measurable coefficients, *Math. USSR Izvestija* **16** (1981), 151-164. English translation in *Izv. Akad. Nauk SSSR* **44** (1980), 81-98.

[6] J. L. Lewis: Regularity of the derivatives of solutions to certain degenerate elliptic equations, *Indiana J. Math.*, **32** (1983), 849-858.

[7] Ibid: Note on a theorem of Wolf, «Holomorphic function theory workshop», Ed. by F. Gehring, MSRI.

[8] J. MANFREDI and A. WEITSMAN: On the Fatou theorem for p-harmonic functions, *Commun. in P.D.E.,* **13** (1988), 651-668.

[9] J. MOSER: On Harnack's inequality for elliptic differential equations, *Comm. Pure Appl. Math.,* **14** (1961), 577-591.

[10] T. WOLFF: Gap series construction for the p-Laplacian, to appear in the *Annali Scuola Normale Superiore di Pisa, Classe di Scienze.*

CAPILLARY SURFACES: A PARTLY HISTORICAL SURVEY

ROBERT FINN

The study of capillary surfaces leads to challenging mathematical problems in global differential geometry and nonlinear analysis. Much has been achieved in recent years, and a central purpose of this article is to outline some of the lines of attack, the successes and also some failures, so as make current developments accessible in general terms to a general mathematical public. I have tried to emphasize unifying principles where they appear, and I hope the material may have something to offer also to cognoscenti. It seems to me that insight is deepened when ideas are seen in the context of their development, and I have therefore included a brief historical survey. The discussion here should be viewed as an invitation to – not a replacement for – more complete material, such as the new historical treatise by Emmer and Tamanini [23]. Similarly, in the interest of providing an overview, I have suppressed details of the mathematical procedures. For a fuller understading the reader will want to consult the books [34, 50, 71, 80] and, of course, the original literature.

From a physical point of view, a capillary surface is the interface that appears whenever two liquids (or liquid and gas) are situated adjacent to each other and do not mix. Abundant examples are available from daily experience: spray from a waterfall, a drop of morning dew on a plant leaf, raindrops on a window pane, the upper surface of a lake (with or without waves), a drop of one liquid (or gas) floating on (or in) another, suspended drops from a pipette, the curved upper surface in a cylindrical capillary tube (or soda straw), a liquid bridge joining two rigid surfaces.

All these configurations are characterized according to a common and simply stated geometrical requirement on their mean curvature, under a geometrical boundary condition. It is useful to see how such a unifying perception came about.

1. ANTIQUITY THROUGH AGE OF ENLIGHTENMENT

The earliest written cognizance that something special happens at a surface interface seems to be that of Aristotle (c. 350 B.C.) who wrote [2] that a broad body – even of heavy material such as iron or lead – would float on a water surface, while a narrow tapered body such as a needle – even if of lighter material – would sink. In fact, only the first part of the statement is correct; inconsistencies in the reasoning to support the second part were later pointed out by Galileo [40].

Leonardo da Vinci wrote in 1490 on the physical mechanism for the formation of liquid drops. He concluded (correctly) that there must be forces analogous to but distinct from the gravitational force. Since the calculus had not yet been invented, Leonardo was not in a position to give quantitative estimates for these forces; such estimates were first obtained several centuries later by Laplace, Gauss, van der Waals and others.

Although no satisfactory general theory appeared prior to 1800, capillarity remained a topic of active interest in the leading scientific circles, and numerous experiments were made and theories postulated. In Fig. 1 is reproduced a page [95] from the Philosophical Transactions of the Royal Society, 1712, in which Brook Taylor describes the configuration that resulted (upper right of Fig. 2) when a wedge formed by two vertical glass plates was placed in a water reservoir; he commented that the traces on the plates of the free surface «come very near to the common hyperbola».

The experiment was repeated by Hauksbee [55] with a coordinate grid on one of the plates, as shown in Fig. 2. Hauksbee also used other liquids, including «spirits of wine».

Segner [82] attempted to relate the rise height in a capillarity tube to the curvature of the free surface. He was however not clear on how the curvature should be defined, and he was led to erroneous results. In this respect a decisive step – not only for capillarity theory but also for the differential geometry of surfaces – was taken by Thomas Young [107] in 1805. Young introduced the notion of *mean curvature* H of a surface, and showed – at least in particular situations – *that* H *is proportional to the pressure change across a capillary surface.* Young gave further a heuristic reasoning to show that *the angle* γ, *formed between the free surface* S *and the walls of a rigid boundary* Σ *that supports the fluid configuration, is a physical constant depending only on the materials* (and not on the shape of S or of Σ, on the amount of fluid or on the thickness of the bounding walls). Putting this information together and applying the laws of hydrostatics, he obtained the first reasonably correct estimate for the rise height in a narrow circular capillary tube.

It is remarkable that Young did all that without the use of formulas or of mathematical symbols, which he professed to disdain. One consequence of this ego-

IX. *Part of a Letter from Mr.* Brook Taylor, *F. R. S. to Dr.* Hans Sloane *R. S. Secr. Concerning the Aſcent of Water between two Glaſs Planes.*

THE following Experiment ſeeming to be of uſe, in diſcovering the Proportions of the Attractions of Fluids, I ſhall not forbear giving an Account of it; tho' I have not here Conveniencies to make it in ſo ſucceſsful a manner, as I could wiſh.

I faſten'd two pieces of Glaſs together, as flat as I could get; ſo that they were inclined in an Angle of about 2 Degrees and a half. Then I ſet them in Water, with the contiguous Edges perpendicular. The upper part of the Water, by riſing between them, made this *Hyperbola*; [See *Fig.* 5.] which is as I copied it from the Glaſs.

I have examined it as well as I can, and it ſeems to approach very near to the common *Hyperbola.* But my *Apparatus* was not nice enough to diſcover this exactly.

The Perpendicular *Aſſymptote* was exactly determined by the Edge of the Glaſs; but the Horizontal one I could not ſo well diſcover. I am,

Sir,

Bifrons near *Canterbury,* June 25. 1712

Your moſt humble Servant,

BROOK TAYLOR.

Fig. 1. From Phil. Trans. Roy. Soc. London, 1712.

centricity is that his results are often expressed in such cumbersome language as to defy comprehension.

In 1806 Laplace [63] introduced independently the notion of mean curvature, and obtained for it a formal analytical expression. Using again the constancy of contact angle (for which he did not give a satisfactory proof but which could be inferred in this case from symmetry) he obtained his celebrated approximation for

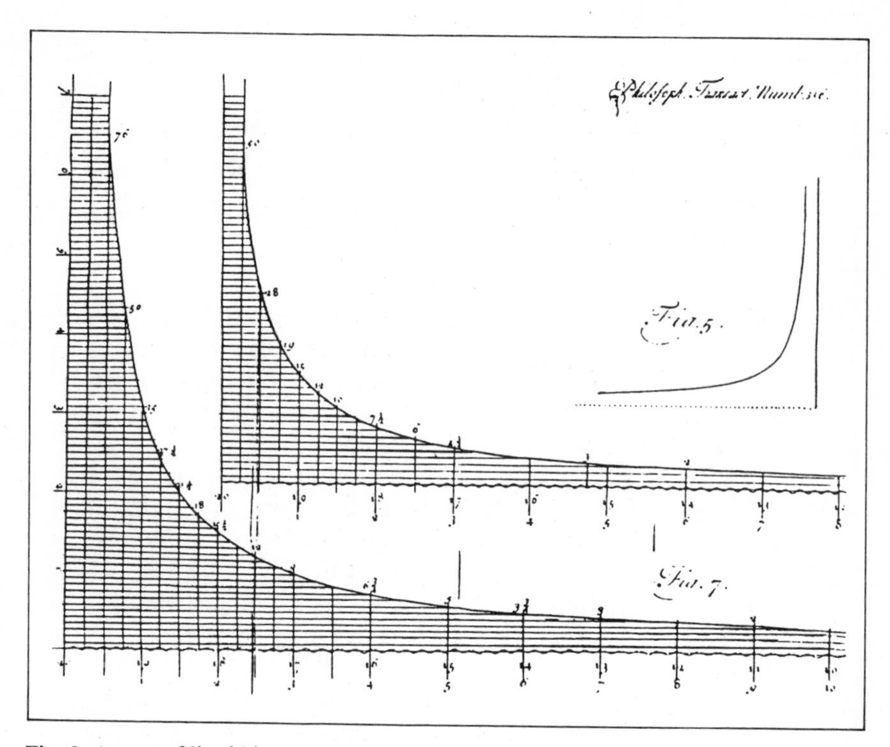

Fig. 2. Ascent of liquid in a corner.

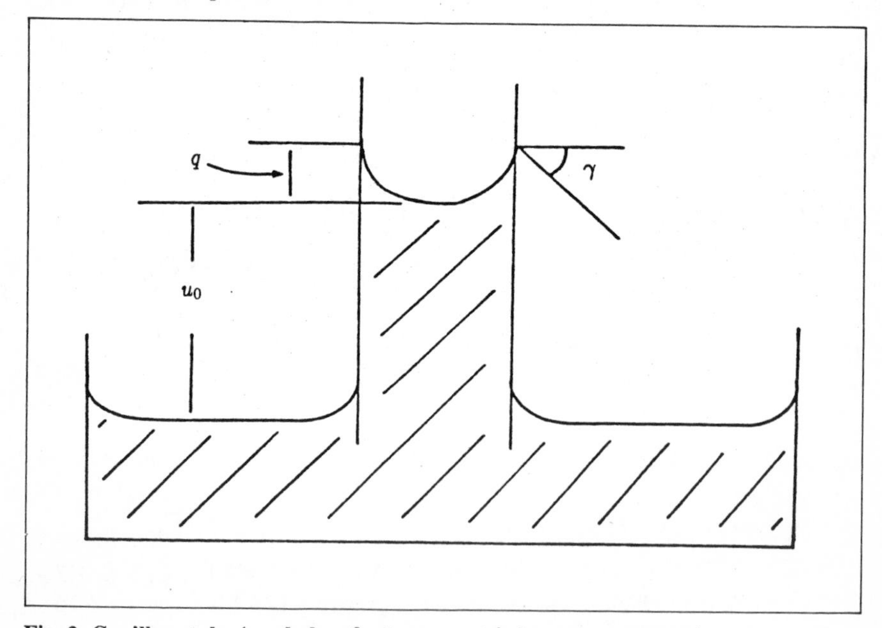

Fig. 3. Capillary tube (symbols refer to symmetric homogeneous case).

the rise height u_0 at the center of a (narrow) circular tube of radius a in a large fluid reservoir (see Fig. 3)

$$u_0 \sim \mathcal{L}(B;\gamma) \equiv \frac{2\cos\gamma}{\kappa a} - \left(\frac{1}{\cos\gamma} + \frac{2}{3}\frac{1-\sin^3\gamma}{\cos^3\gamma}\right)a. \tag{1}$$

Here $\kappa = pg/\sigma$, $\rho =$ density change across free surface, $g =$ gravitational acceleration, $\sigma =$ surface tension. $B = \kappa a^2$, see III.

As one consequence of his formula for mean curvature, Laplace derived the equation

$$\operatorname{div} Tu = \kappa u + \lambda, \quad Tu = \frac{\nabla u}{\sqrt{1+|\nabla u|^2}}, \qquad \kappa, \lambda = \text{const}\,. \tag{2}$$

for the height of a capillary surface $u(x,y)$ in a gravity field. Here κ must be taken negative if the heavier fluid lies above the free surface, as in a drop from a pipette; λ is a Lagrange parameter that appears if there is a volume constraint, and is in general determined by the constraint.

Prior to the work of Young and of Laplace, Johann Bernoulli proposed his universal Principle of Virtual Work, which was to provide an effective new weapon for the study of capillary surfaces and of their stability. According to this principle, *a system of particles is in mechanical equilibrium if the first variation of its mechanical energy vanishes for any (virtual) displacement consistent with the contraints.*

II. PRE-MODERN TIMES, 1800-1950

It was Gauss [41] who took the basic step of applying to the problem the Bernoulli principle. He obtained from it at once, both the differential equation (2) and also the boundary condition on the contact angle. More generally, one obtains by this procedure the relation, for any external force field with potential Υ per unit mass

$$2H = \frac{\rho}{\sigma}\Upsilon + \lambda \tag{3}$$

at each point on the free surface interface S, while for the angle γ that S makes with a rigid supporting surface Σ at a triple interface one finds

$$\cos\gamma = \beta \tag{4}$$

where β is the local relative adhesion coefficient, depending only on the materials, see e.g. [34], Chapter 1.

Fig. 4. Web of diadema spider.

Gauss proceeded from detailed hypotheses on the nature of the local attractive forces in order to derive the relevant energy expressions; there may be some question about the procedure, as recent studies have shown that the forces must be expected to act only within molecular distances, whereas Gauss assumed a continuous mass distribution. However, one can proceed alternatively with the macroscopic assumption that an energy is associated with each surface interface, depending only on the materials and proportional to the area of the interface, see [34], Chapter 1.

The results of Gauss put the entire theory onto a conceptually sound basis: *capillary surfaces are characterized as stationary points for an energy functional; stable surfaces must be local minima.* This step is basic for much of the ensuing theory.

The energy criterion was applied by Lohnstein [69] to study the stability of drops from a pipette (the prediction of the breaking point led to procedures for measuring surface tension, also for metering out small precisely controlled quantities of medicines). Lohnstein took as stability limit (without proof) the criterion that the

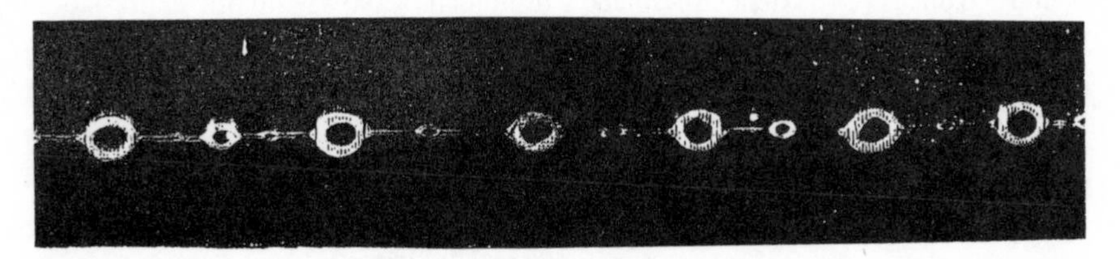

Fig. 5. Photomicrograph from circular strand of Figure 4.

volume of the drop below the nozzle should achieve a local maximum when expressed in terms of height change from tip to nozzle. This condition was used also by later authors and was verified formally by Wente [106].

Plateau and later Rayleigh [81] considered an infinite tube of liquid of circular section (which yields an exact solution in zero gravity) and showed that *it admits (infinitesimal) periodic volume preserving energy reducing deformations, with period L equal to the circumference C.* This discovery was not entirely new with these authors; it was known long before their birth and used to advantage by every spider. Fig. 4 shows a typical web woven by the common diadema (garden) spider; Fig. 5 is a photomicrograph of a portion of a circumferential strand. Ms. Diadema weaves a solid thread surrounded by fluid, which becomes unstable with the Plateau-Rayleigh period and decomposes into sticky balls (several hundred thousand per web).The radial strands (along which the spider moves to reach her prey) are of solid thread, without fluid. Figs. 4 and 5 are taken from the classic book of Boys [8].

Subsequent to the work of Rayleigh, capillarity studies suffered a quietus of half a century. The subject fell into disfavor, few papers were written, and little significant progress was made. A notable exception was the paper [7] in which the nondimensional «Bond number», now universally used, was introduced as a measure of size of a configuration. It may be worth noting that this paper was published with a disclaimer, that it was contributed by the authors and that the editors were not responsible for its content.

III. MODERN TIMES, 1950 +

The formula (1) of Laplace remained the single available known quantitative relation for rise height in a capillary tube for over 150 years, and was universally used in engineering work. This despite the fact that Laplace had not proved the relation, much less provide information on its range of applicability. The right side of (1) becomes negative if $B = \kappa a^2 > 8$, and thus provides no information in this range. Laplace gave also estimates for large B, again without proof or error bounds. Concus [14] extended Laplace's results and gave asymptotic expressions for the entire solution traverse, and numerically derived error bounds. The first step toward providing rigorous analytical information was taken by Siegel [85] in 1980, who proved the asymptotic correctness of (1) for any fixed γ as $B \to 0$. The problem was then taken up in [28], where the asymptotic estimate

$$\mathcal{L}(B;\gamma) < u_0 < \mathcal{L}(B;\gamma) + \frac{\cos\gamma(1+2\sin\gamma)}{6(1+\sin\gamma)^4}B + 0(B^2) \tag{5}$$

is obtained (a strict upper bound is given in implicit form), and again in the Russian edition of [34], where the strict bound, for $0 < B < 6$,

$$\mathcal{L}(B;\gamma) < u_0 < \mathcal{L}(B;\gamma) + \frac{\cos\gamma(1+2\sin\gamma)}{6(1+\sin\gamma)^4}B \tag{6}$$

is established. In the spirit of the observation of Bond and Newton (see above) the relations (5) and (6) are written nondimensionally, and u_0 denotes the ratio of center height to the radius a. Bounds are given also for the height u_1 on the boundary wall, and the idea was then extended by Siegel [87], who set

$$f(\gamma) = \frac{1-\sin\gamma}{\cos^2\gamma}\left(\frac{1}{3}\frac{1-\sin^3\gamma}{\cos^3\gamma} - \frac{1}{2}\frac{\sin\gamma}{\cos\gamma}\right)$$

$$\mathcal{S}(B;\ \gamma;r) = 2\frac{\cos\gamma}{B} + \frac{2}{3}\frac{1-\sin^3\gamma}{\cos^3\gamma} - \sqrt{\frac{1}{\cos^2\gamma} - r^2}$$

and found

$$\mathcal{S}(B;\gamma;r) - f(\gamma)B < u(r) < \mathcal{S}(B;\gamma;r) + f(\gamma)B \tag{7}$$

for the (nondimensional) height $u(r)$ throughout the trajectory $0 \le r \le 1$.

Some of the above results are valid for every B; all are asymptotically exact as $B \to 0$, but inexact for large B. Nevertheless, *the method can be extended to yield explicit meaningful bounds for every* B, see [26] or [34] Chapter 2. As an example, we show in Fig. 6 strict upper and lower bounds (solid lines) for the non-dimensional meniscus height $q = u_1 - u_0$, as compared with earlier estimates based on formal expansions designed for small B (inner) or large B (outer).

The sessile liquid drop, supported by a horizontal plane of homogeneous material, is governed by (essentially) the same equation as that for the circular capillary tube; however the radius of the contact circle is no longer known apriori, also the surface need not be a graph (see Fig. 7). The symmetry of those surfaces that are

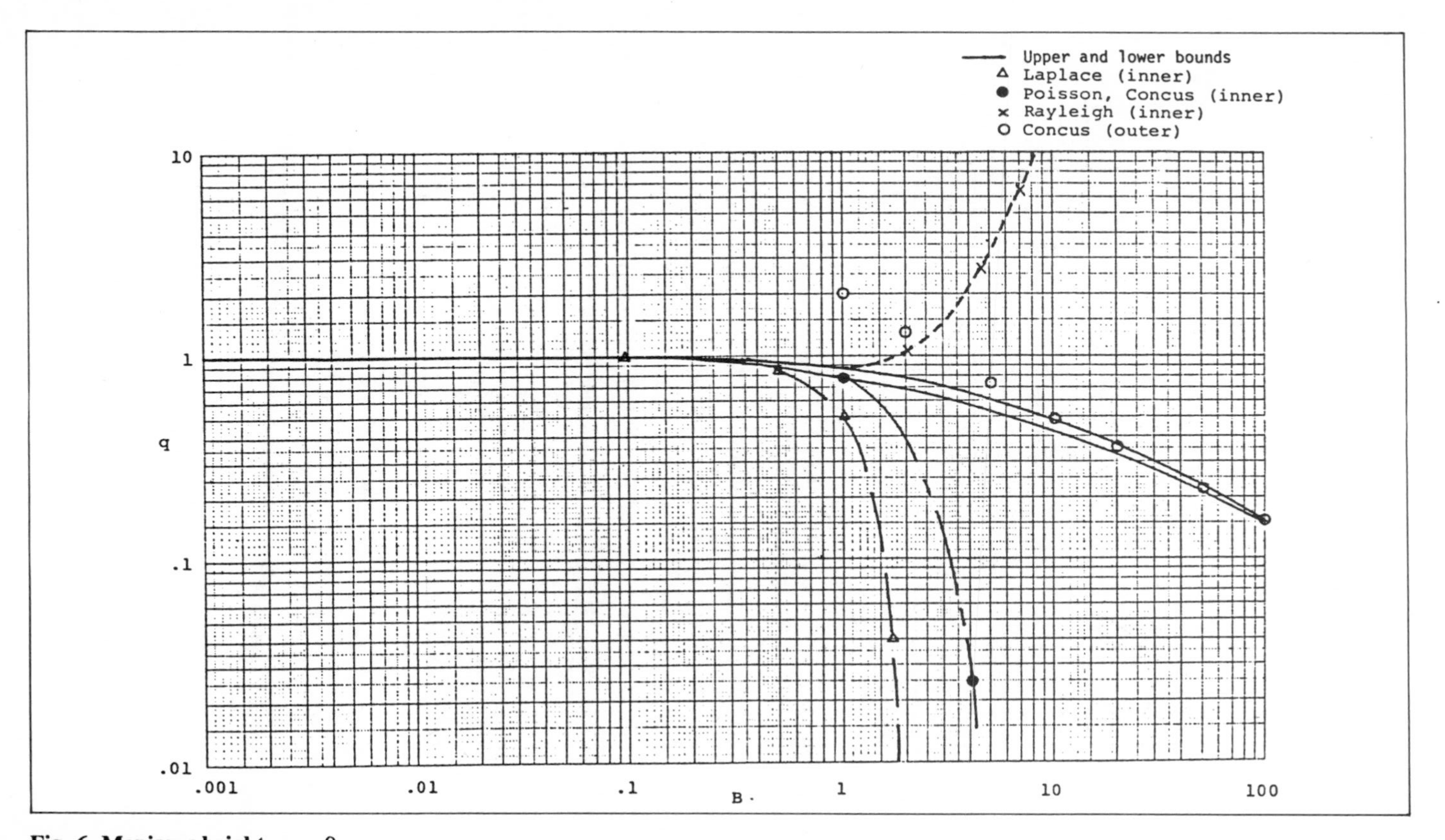

Fig. 6. Meniscus height; $\gamma = 0_{-}$.

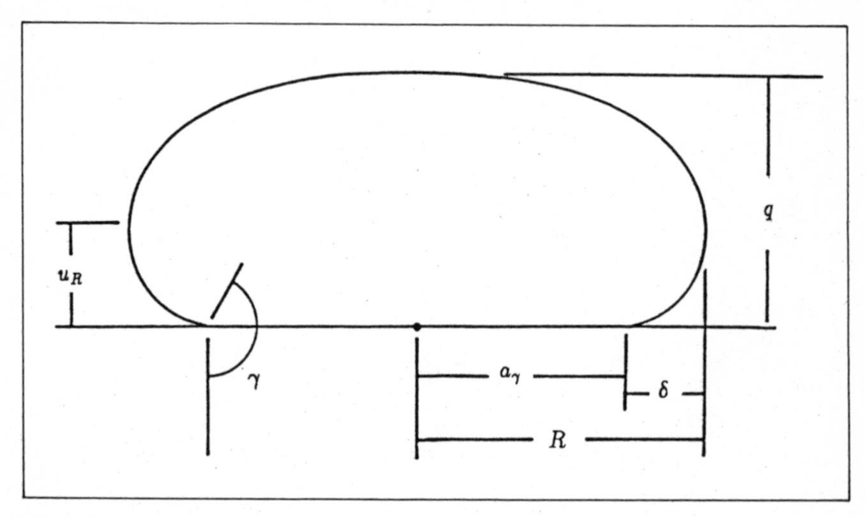

Fig. 7. Sessile drop.

graphs was proved by Serrin [83]; the proof was extended to the general case by Wente [105].

Size and shape estimates analogous to those for capillary tubes continue to hold, although the proofs become somewhat more intricate [29]. Of particular interest is a *nonuniformity in asymptotic behaviour as volume* $V \to 0$, *depending on whether or not* $\gamma = \pi$. Letting P be the radius of a ball of volume V and setting a = radius of wetted disk, we set $\mathcal{B} = \kappa P^2, B = \kappa a^2$ and find that as $\mathcal{B} \to 0$,

$$B \sim \frac{\sin^2 \gamma}{s^2(\gamma)} \mathcal{B}, \quad \gamma \neq \pi \tag{8}$$
$$B \sim \frac{2}{3} \mathcal{B}^2, \quad \gamma = \pi.$$

Here $s^3 = \frac{3}{4} \int_0^\gamma \sin^3 \theta d\theta$.

Thus, *if* γ *is close to* π, *a small drop tends to sit on a point, and can be expected to exhibit instability with respect to a kind of «rolling» motion.* The effect is illustrated in Fig. 8; note that the ordinate for the $\gamma = 5\pi/6$ curve had to be multiplied by 10 to keep it on scale.

Another point of interest is the «overhang» δ that occurs when $\gamma > \pi/2$ (see Fig. 7). *For* $a \to \infty$ *we find the explicit value* [29].

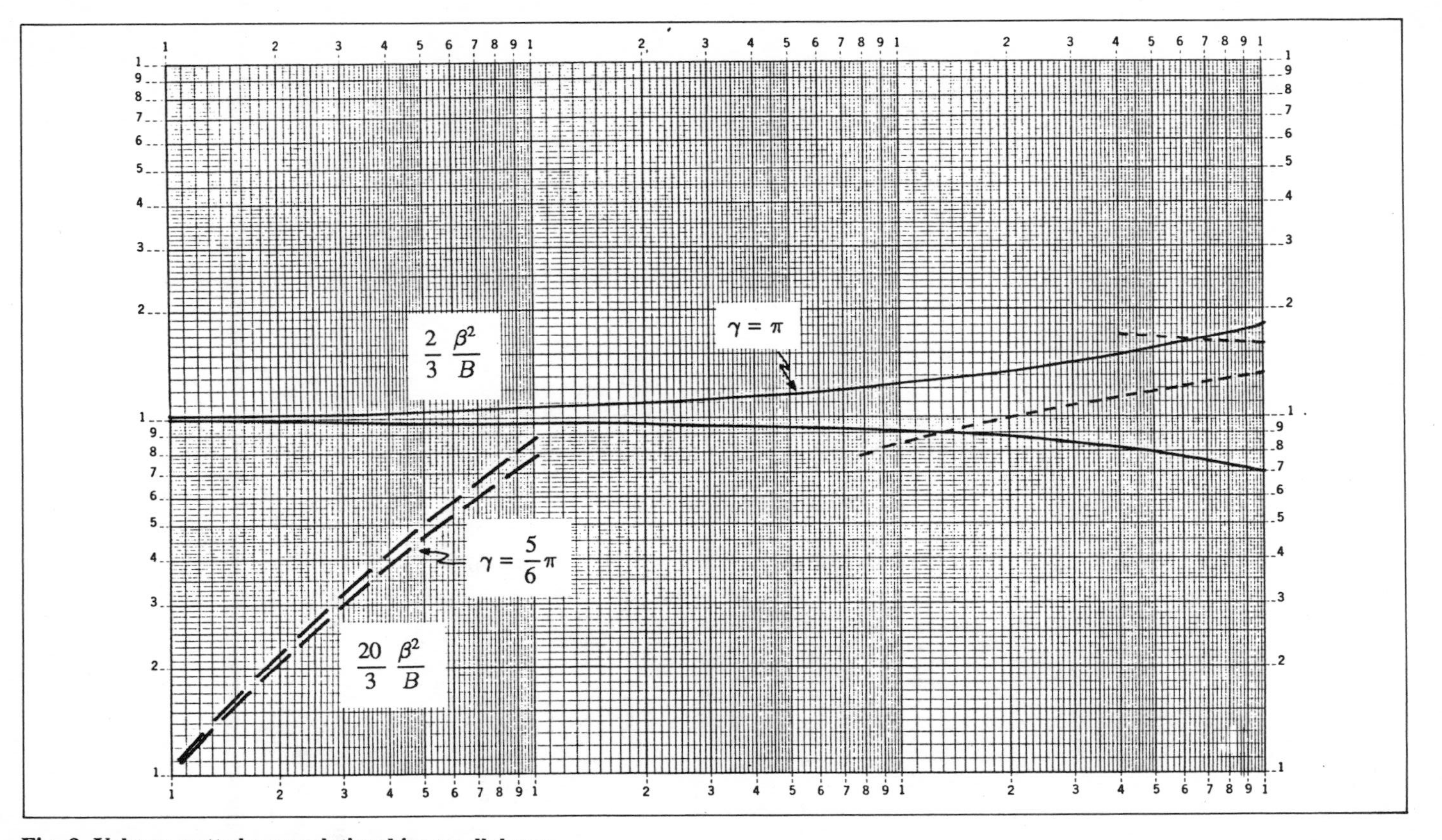

Fig. 8. Volume-wetted area relationship; small drops.

$$\lim_{a\to\infty} \sqrt{\kappa}\delta = \sqrt{2} - \log(1+\sqrt{2}) - 2\cos\frac{\gamma}{2} + \log\frac{1+\cos\frac{\gamma}{2}}{\sin\frac{\gamma}{2}}. \tag{9}$$

It should be noted that *the height of a drop does not increase monotonely with its volume.* The controlling parameter is (at least when $0 \le \gamma \le \pi/2$) the value $k = \kappa^{3/2}V/\sin\gamma$. In [26] is shown *the existence of positive* k_1, k_2 *such that the maximum height is always attained for a value* k *with* $k_1 < k < k_2$, see also [34], Chapter 3.

Sessile drops are strictly and globally stable, in the sense that the energy is minimized under the constraint and boundary condition, see Gonzalez [51].

For a drop pendent from a ceiling (or from a nozzle) the sign of κ in (2) must be reversed, and the behaviour of the solutions can be very different. Kelvin was the first to recognize the existence of global solutions (of an associated parametric problem) that are not physically observed, see Figure 9. These solutions are important, at least in the sense that *the lower tip will appear as a (stable) pendent drop,* see e.g., the discussion in [34], Chapter 4, or in the Appendix to the later Russian edition. It is shown in [19] that *solutions with arbitrarily many bulges can be obtained,* and conjectured (and partly proved) that these solutions converge uniformly in compacta to a singular solution, whose existence is shown in [17] and in [5]. In [32] a priori bounds are given for the maximum bulge width, volume and other drop parameters.

Stability questions for pendent drops have been studied for over a century, by many authors. We have already referred to the work of Lohnstein. The most precise and extensive of the known results are due to Wente ([106] and Appendix noted above) who was also the first to distinguish clearly the cases arising from varying boundary conditions and constraints. For the case of fixed aperture and constant volume, Wente showed *the existence of stable drops with both neck and bulge.* Fig. 10 shows a «kitchen sink» experiment of a drop of water in a reservoir of castor oil (slightly less density), illustrating the phenomenon.

The stability of a liquid bridge joining parallel plates (Fig. 11) was studied by Athenassenas [3] and by Vogel [102], [103]. *For the particular case* $\gamma = \pi/2$ *they obtained the criterion* $L < \frac{1}{2}C$ *for stability.* The factor $\frac{1}{2}$, which does not appear in the Rayleigh criterion mentioned earlier, arises from the boundary condition, which allows deformation with half a Rayleigh period.

Let us return to the experiments of Taylor and of Hauksbee for fluid in a wedge domain. A number of «proofs» of hyperbolic trace appear in the literature; however,

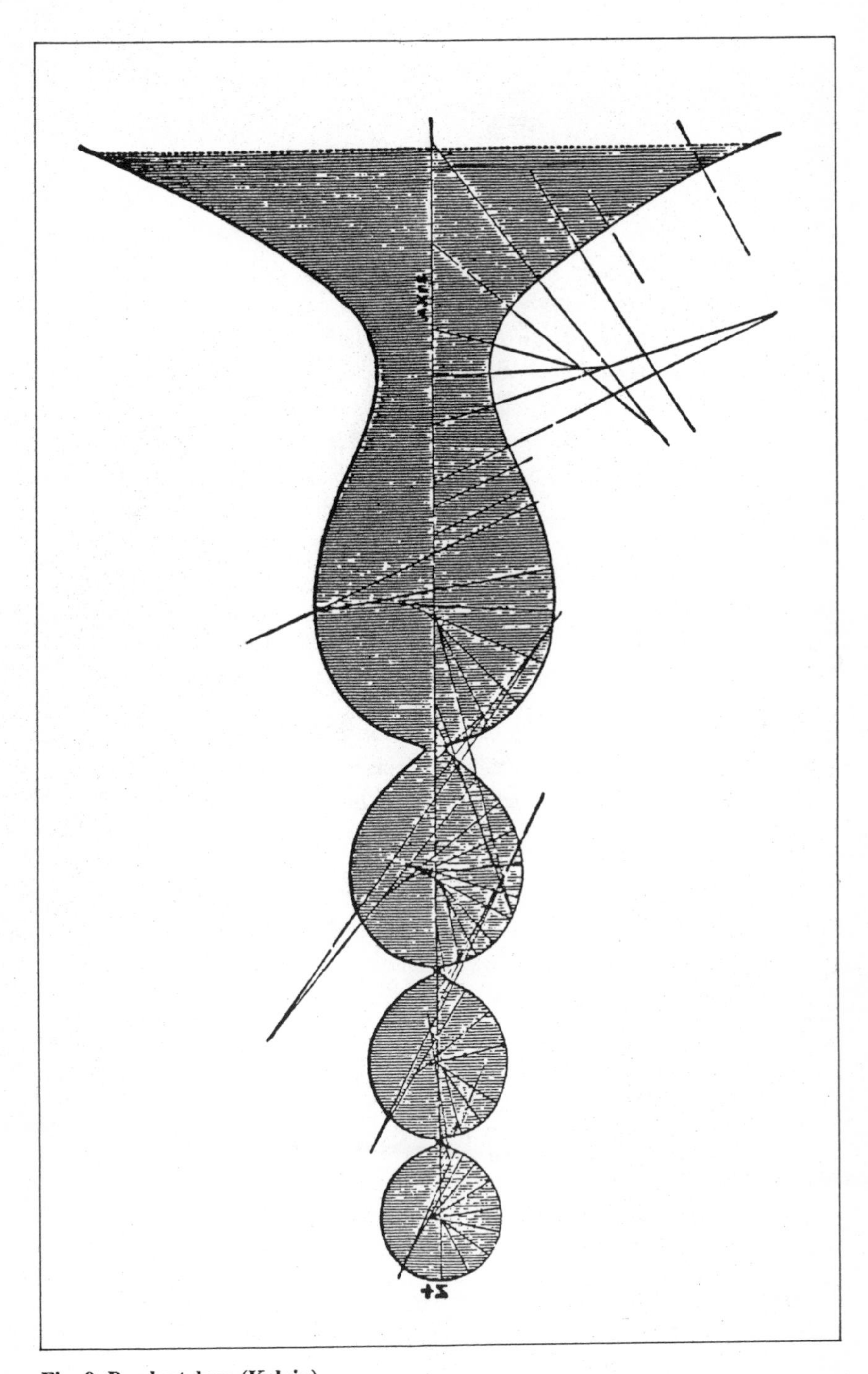

Fig. 9. Pendent drop (Kelvin).

Fig. 10. Stable pendent drop with neck and bulge.

the actual behaviour can be quite different. It is shown in [16] that if 2α is the wedge opening and if the wedge includes a domain Δ as indicated in Fig. 12, then *in a «positive» gravity field the height* u *of the free surface satisfies*

$$u \sim \frac{\cos\theta - \sqrt{k^2 - \sin^2\theta}}{k\kappa r}, \quad k = \frac{\sin\alpha}{\cos\gamma}, \tag{10}$$

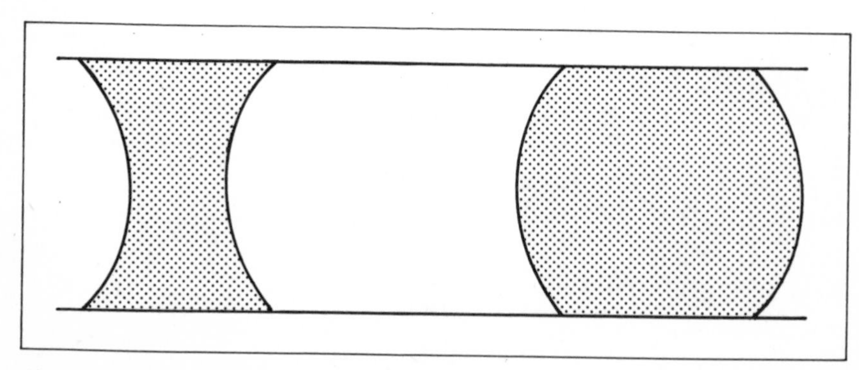

Fig. 11. Liquid bridge between parallel plates.

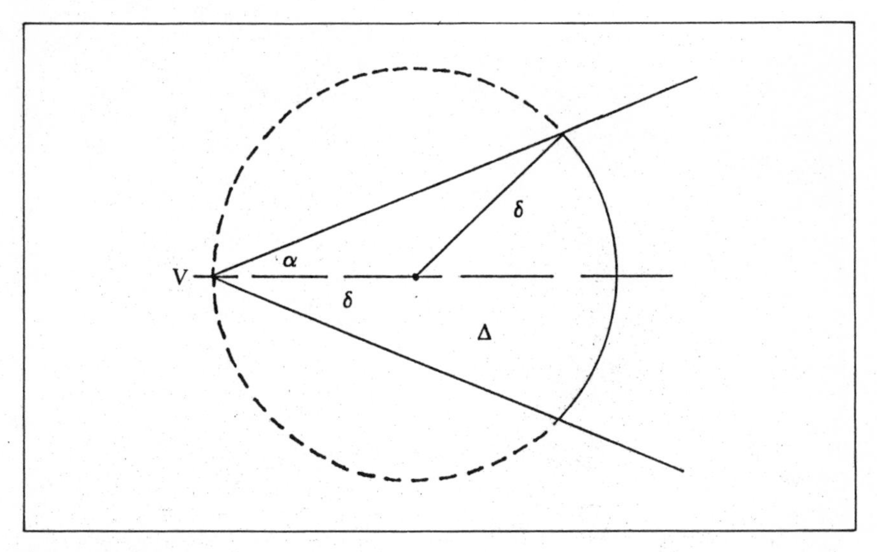

Fig. 12. Wedge domain.

if $0 \leq \gamma < \pi/2, \alpha + \gamma < \pi/2$, *in the sense that the difference of the two sides is bounded.* Here r, θ are polar coordinates based at the vertex. Thus in this case the hyperbolic behaviour is verified. But *if* $\alpha + \gamma \geq \pi/2$ *there holds*

$$(11) \qquad 0 < u \leq \frac{2}{\kappa\delta} + \delta$$

thoughout Δ, which gives a quite different behaviour. If one starts with a large γ and then decreases the contact angle until $\frac{\pi}{2} - \alpha$ is reached, (11) continues to hold. But by (10), for all smaller $\gamma, u \to \infty$ at V. Thus, *the solutions depend discontinuously on the prescribed data.* This behaviour was verified experimentally by Tim Coburn in the Medical School at Stanford University, using two plates of acrylic plastic with distilled water. The result, with the angle α varied by about $2°$, is shown in Fig. 13, and establishes the contact angle of water with acrylic plastic to be in the range $78°$-$81°$. A carefully controlled laboratory experiment would presumably yield still greater accuarcy.

The corner behaviour has been studied further by a number of authors. Simon [88] showed that *if* $0 < \gamma \leq \pi/2, \alpha < \pi/2, \alpha + \gamma > \pi/2$, *then* ∇u *is continuous up to* V. Lieberman [68] and Miersemann [73] showed *that* ∇u *is in fact Hölder continuous,* and Miersemann [74] obtained a complete asymptotic expansion for u. Tam [91] showed that *if* $0 < \gamma \leq \pi/2, \alpha < \pi/2$ *and* $\alpha + \gamma \geq \pi/2$, *then the surface normal is continuous up to* V. If $\alpha + \gamma < \pi/2$, it seems likely that the

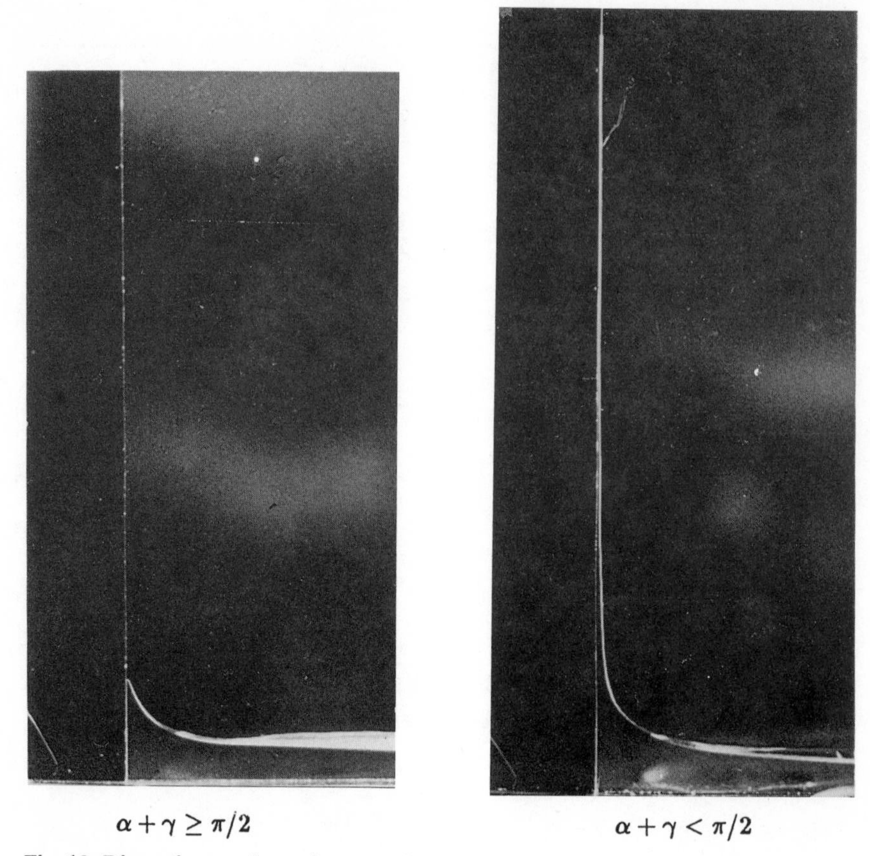

$\alpha+\gamma \geq \pi/2$ $\alpha+\gamma < \pi/2$

Fig. 13. Discontinuous dependence on data: $g > 0$.

vertical component N_3 of the normal will $\to 0$ at V, but this has not yet been proved.

Korevaar [57] obtained the striking result that *if* $\alpha > \pi/2$ *and* $\gamma \neq \pi/2$ *then there exist solutions that are discontinuous at* V. Thus, even though the bound (11) continues to hold in this case, there can be no bound on $|\nabla u|$.

Miersemann [75] improved the estimate (10) by showing that in fact *there exists* $\epsilon > 0$ *such that*

$$\left| u - \frac{\cos\theta\sqrt{k^2 - \sin^2\theta}}{k\kappa r} \right| < Cr^{\epsilon} \tag{12}$$

near V, *when* $\alpha + \gamma < \pi/2$.

The above results would be vacuous if solutions did not exist under the indicated conditions. The first existence theorem for the capillary tube of general section

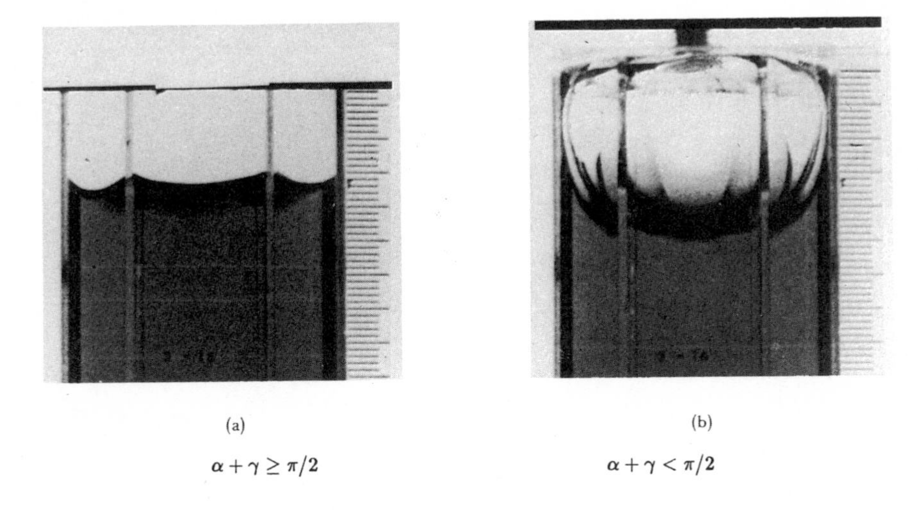

(a) $\alpha+\gamma \geq \pi/2$

(b) $\alpha+\gamma < \pi/2$

Fig. 14. Discountinuous dependence on data: $g = 0$.

is due to Emmer [22], who considered Lipschitz domains and used a variational procedure of minimizing energy. He was led to the requirement $L < \tan\gamma$ for the Lipschitz constant L. This condition corresponds exactly to $\alpha + \gamma > \pi/2$ at an isolated corner, and must be expected, as the surface determined by the right side of (10) has infinite area, yields infinite wetted surface on the bounding walls, and infinite gravitational energy.

The existence of a solution at an isolated corner when $\alpha + \gamma \leq \pi/2$ is shown in [37]. This paper additionally covers the case $\gamma = 0$ for a smooth boundary, which is also excluded by Emmer's condition. For further existence theorems and for discussion of boundary regularity, see the references cited in [34].

The case of zero gravity ($\kappa = 0$) presents another world and is not accessible to Emmer's methods. In this case, it can be shown that *if $\alpha+\gamma < \pi/2$ at a corner, there is no solution as a graph over the domain [15]. But for any inscribable polygon, if $\alpha+\gamma \geq \pi/2$ at each vertex, an explicit solution can be achieved as a lower hemisphere whose equatorial circle is concentric with the inscribed circle.* Thus, the discontinuous behaviour persists and in fact becomes more pronounced. Fig. 14 shows verification of this behaviour by W. Masica in a drop tower experiment using identical hexagonal cylinders, with different fluids. If $\alpha+\gamma < \pi/2$ the fluid climbs into the corner and attemps to fill out an entire cylindrical neighborhood of a vertical edge. Thus, even in the idealized case of a semi-infinite cylinder, the free surface would not cover the entire base domain if the total volume of fluid is finite.

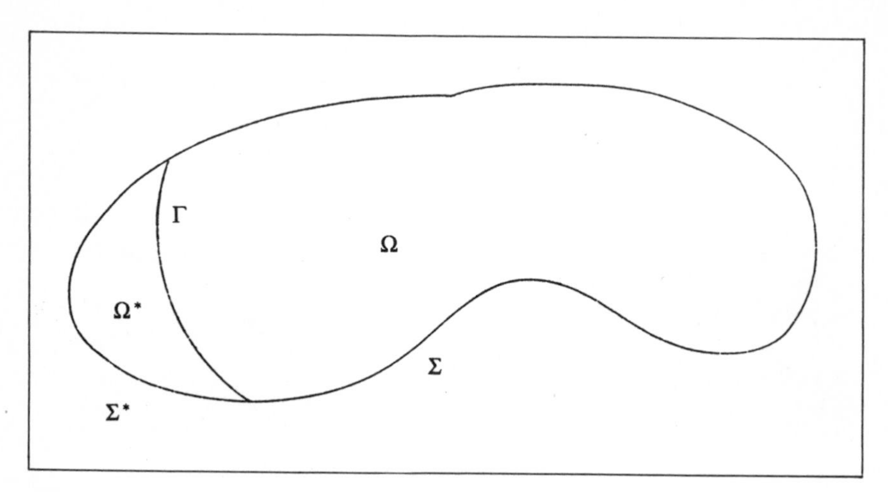

Fig. 15. Existence Criterion.

A general necessary condition for existence of a zero gravity solution was first formulated in [15]. Giusti [46] introduced the condition independently and found it under some restrictions to be also sufficient. The ideas were developed further in [31], where a «subsidiary variational problem» in one lower dimension is introduced, and it is shown that *the original problem has a solution if and only if there is no minimizing solution for the subsidiary problem.* In specific terms, *let* Ω *denote the section of the tube,* $\Sigma = \partial\Omega$, *and set* $R_\gamma = |\Omega|/|\Sigma|\cos\gamma$. *There exists a capillary surface as a graph over* Ω *if and only if for every subarc* Γ *of a semicircle of radius* R_γ *that lies in* Ω *and meets* Σ *in angles* γ *there holds* (See Fig. 15)

$$|\Gamma| - |\Sigma^*|\cos\gamma + \frac{1}{R_\gamma}|\Omega^*| > 0\,. \tag{13}$$

In general there will be only a finite number of arcs Γ satisfying the requirements, so the condition can be verified in a finite number of steps. A case of particular interest occurs when there is no such arc Γ, so that the condition is satisfied vacuously. An example of such a case is the configuration of Fig. 16, bounded by two parallel lines and two semicircles. Here there is no admissible Γ for any $\gamma \neq \pi/2$, and we conclude that *regardless of* h, *a solution always exists* (if $\gamma = \pi/2$ the functional is clearly positive, and in fact any constant function yields a solution). This situation contrasts sharply with that of an ellipse (Fig. 17), for which *if* $\gamma \neq \pi/2$ *the criterion precludes the existence of a solution for any large enough eccentricity, depending on* γ.

Fig. 16. Solution exists, any γ.

The question of boundary regularity has a special interest. It can be shown (cf the references in [34]) that *if* $0 < \gamma < \pi$, *if* Σ *is smooth and if a bounded solution* $u(x)$ *exists, then* $u(x)$ *is smooth up to* Σ *and behaves there like a solution of the classical Neumann problem.* Such behaviour cannot be expected when $\gamma = 0$ or π, as then $|\nabla u|$ necessarily becomes unbounded near Σ. In [33] is shown that *in these cases there always holds*

$$|\nabla u| > Cd^{-1/2} \tag{14}$$

where d *is distance to* Σ. *Further, for the angle* θ *between* ∇u *and the (extended) surface normal* v, *there holds*

$$|\theta| < Cd^{1/2}. \tag{15}$$

The exponent here is sharp, and the constants C *admit the simple choices* $C = \frac{1}{2} - \epsilon$ *in (14),* $C = 2 + \epsilon$ *in (15) under certain conditions. A consequence of (14),*

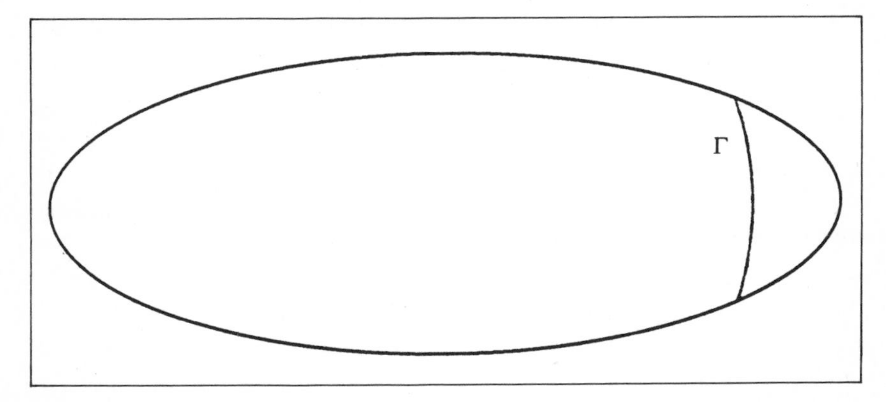

Fig. 17. Solution fails to exists.

(15) is that the boundary condition $v \cdot Tu = \pm 1$ is achieved strictly in the sense

$$|v \cdot Tu \mp 1| < Cd \tag{16}$$

with $C = 4 + \epsilon$ under the particular conditions referred to.

When are capillary surfaces convex? *For a cylindrical capillary tube, it does not suffice in general for the section Ω to be convex;* a counterexample appears in [30]. However, *if in addition $\gamma = 0$, then convexity follows.* This was proved by Chen and Huang [12] for zero gravity and by Korevaar [59] for $g > 0$.

Chen [13] showed that *if $g = 0$ and Ω is convex, then the minimum point is uniquely determined, for any $\gamma, 0 \leq \gamma < \pi/2$.* The reasoning was later improved by Siegel, who also extended the result to non zero gravity. Huang [56] showed that *the minimum point has distance at least $|\Omega| / |\Sigma|$ from $\Sigma = \partial\Omega$.*

Finn has shown by example that a *sessile drop with convex wetted area on a support plane need not itself be convex.*

M. Miranda has raised the question: if one capillary tube lies strictly interior to another one, does it always raise fluid to a larger height over its section than does the larger tube over the same section? A negative answer was given in [25], where it was also shown that *the outer tube can even raise a larger volume of fluid over the inner section.* But *if the outer tube is circular, the answer is positive [25].*

Siegel [85] showed that *if the inner tube is circular and can contact the outer one at each boundary point from within the domain, then the answer is again positive.*

We turn now to the question of uniqueness. As of this writing uniqueness theorems (for prescribed fluid volume) appear in the literature in exactly two cases:

(i) a cylindrical semi-infinite capillary tube of general section, and (ii) the sessile drop on a horizontal plane. If the plane is deformed by continuous motion into a (semi infinite) tube, existence continues to hold (see Giusti [49]) but uniqueness may fail. As an example, consider the rotationally symmetric tube with vertical section, obtained by adjoining a circular cylinder to a right circular cone, as indicated in Fig. 18. If $\gamma = 45°$ and if the conical part of the tube is continuously filled with increasing amounts of liquid, one obtains always a horizontal surface as shown, for any gravity. Now start with an infinite amount of liquid filling the tube, and continuously reduce the volume. One obtains a family of congruent surfaces, that cannot be planar in view of the $45°$ contact angle, and which must therefore be convex (see [34], Chapter 2). The smallest volume attainable this way must thus be less than the largest volume attainable from the bottom, and so it is clear that *distinct configurations with identical volumes are possible.*

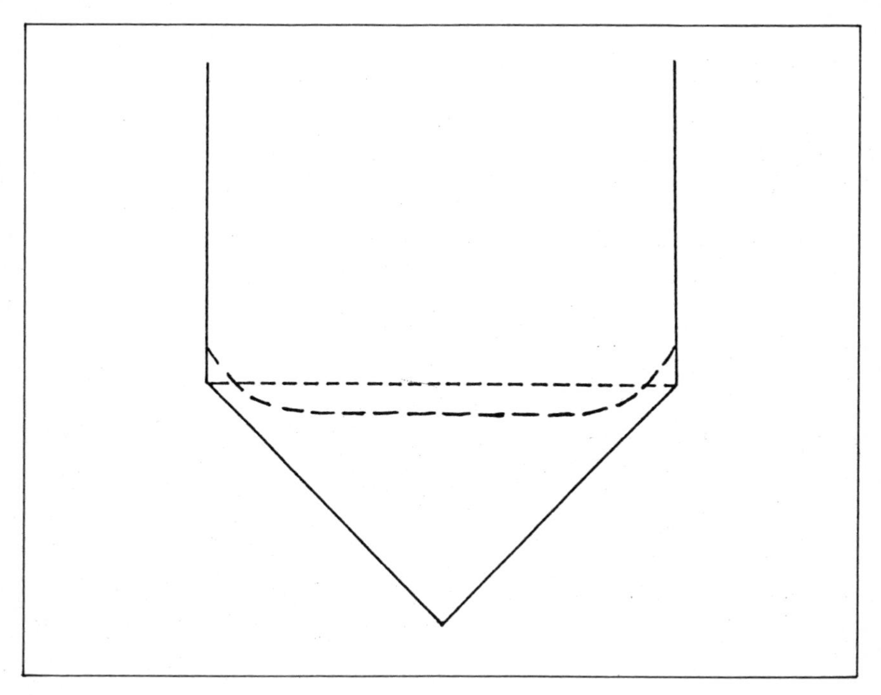

Fig. 18. Nonuniqueness.

It is possible to design the container so that an entire continuum of distinct solutions appears. This was shown in [54] for zero gravity and in [36] for any gravity. These solutions have not only identical volumes and contact angles, but also identical energies, and the question thus arises: which, if any, will be seen physically? In [36] it is shown that if gravity is small enough, the energy will in general not be a local minimum. That result was in turn extended in [21] to any gravity. Further, *it is possible to design the container to be symmetric and arbitrarily close to a closed circular cylinder, so that when half filled with liquid, the zero gravity energy minimizing configuration will be asymmetric.*

We turn our attention finally to some general mathematical questions regarding surfaces $u(x,y)$ of constant mean curvature $H \neq 0$ over a disk $B_R(0)$. We normalize by a coordinate transformation so that $H \equiv 1$. In [24] it was shown that *if* $R \geq 1$ *then* $R = 1$ *and* $u(x,y)$ *describes a lower hemisphere.* In [38] was shown that *there exists* $R_0 = 0.5654062332\ldots$ *and a decreasing* $M(R)$, *with* $M(R_0) = \infty, M(1) = 0$, *such that if* $R > R_0$, *then* $|\nabla u(0)| \leq M(R)$; *thus* $|\nabla u(0)|$ *admits a bound depending only on domain of definition.* Liang [64] has characterized *an increasing* $R^*(R)$, *with* $R^*(R_0) = 0, R^*(1) = 1$, *such that*

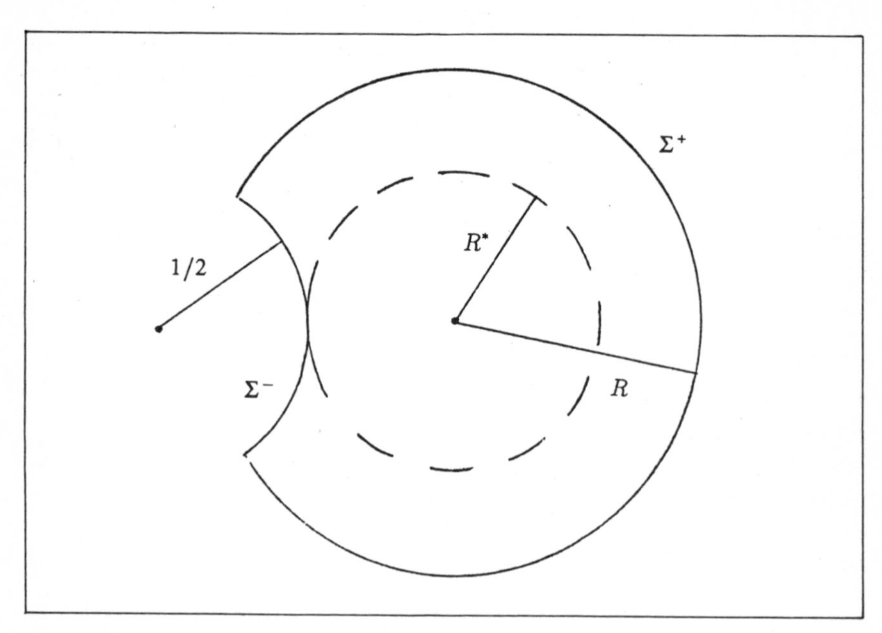

Fig. 19. Moon domain.

$|\nabla u|$ *is bounded in compact subsets of* B_{R^*} *when* $R_0 < R \leq 1$. The proof is based essentially on properties of zero gravity capillary surfaces $v(x, y)$ over a «moon domain» as indicated in Fig. 19, with $\gamma = \pi$ on $\Sigma^-, \gamma = 0$ on Σ^+. The position of the center of Σ^- on the negative real axis is uniquely determined, in order that $v(x, y)$ should exist and have mean curvature $H \equiv 1$. It can then be shown that $|\nabla v(0)|$ majorizes $|\nabla u(0)|$ for any $u(x, y)$ with $H \equiv 1$ in B_R. The disk B_{R^*} is as indicated in the figure; the existence of a bound in B_{R^*} is obtained by an indirect reasoning from properties of the «generalized solutions» introduced by Miranda [77].

The value R^* *is sharp.* However, *if* $|\nabla u|$ *is large in* $B_R \backslash B_{R^*}$, *then* ∇u *must be directed into a narrow segment containing the origin,* see [35, 64].

Chua has recently obtained an explicit estimate for $|\nabla u(0)|$ when $R > R_0$. If $R \leq R_0$ then no gradient bound depending only on domain can hold, but the result then goes over in a natural way into Serrin's Harnack Inequality [84], see [35].

This survey was intended to whet the reader's appetite with a brief general overview of some current activities on capillary surfaces, in the context of historical background. The survey is not complete. The references at the end offer further insight, details of proofs, and leads to material that has not been cited here. We close the smattering that we have offered with the quotation from A. Yu. Davidov,

Theory of Capillary Phenomena, Moscow 1851:

«The outstanding contributions made by Poisson and by Laplace to the mathematical theory of capillary phenomena have completely exhausted this subject and brought it to such a level of perfection that there is hardly anything more to be gained by their further investigation.»

In fairness to the memory of a distinguished scientist, it must be pointed out that the comment was modified for the later French and German summaries of his Treatise.

I wish to thank E. Giusti for informing me of the references to Aristotle and to Galileo.

PARTIAL BIBLIOGRAPHY

[1] F. J. ALMGREN: Existence and regularity almost everywhere of solutions to elliptic variational problems with constraints. *Memoirs A.M.S.* **4** No. 165.

[2] ARISTOTLE: Del Cielo IV (Δ), 6, 313 a-b.

[3] M. ATHANASSENAS: A variational problem for constant mean curvature surfaces with free boundary. *J. Reine Angew. Math.* **377** (1987), 97-107.

[4] E. BAROZZI, M. EMMER and E. H. A. GONZALES: Variational Methods for Equilibrium Problems of Fluids. Preprint, Università de Sassari, to appear.

[5] M.-F. BIDAUT-VERON: Global existence and uniqueness results for singular solutions of the capillarity equation. *Pacific J. Math.* **125** (1986), 317-334.

[6] M.-F. BIDAUT-VERON: New Results Concerning the Singular Solutions of the Capillarity Equation *Variational Methods for Free Surface Interfaces*, Springer-Verlag, 1987.

[7] W. N. BOND and D. A. NEWTON: Bubbles, drops, and Stokes's Law (Paper 2). *Phil. Mag. Ser. 7*, no. 5 (1928), 794-800.

[8] C. V. BOYS: Soap Bubbles, and the Forces Which Mould Them. Society for Promoting Christian Knowledge, London 1902; revised edition, 1916. Reprinted by Dover Publications, New York, 1959.

[9] F. P. BRULOIS: Asymptotic expansions and estimates for the capillary problem. «Proc. Second Int. Coll. Drops and Bubbles». Jet Propulsion Lab., Cal. Inst. Tech., (1982), 344-349.

[10] F. P. BRULOIS: The Limit of Stability of Axisymmetric Rotating Drops. *Variational Methods for Free Surface Interface*, Springer-Verlag, 1987.

[11] J.-T. CHEN: On the existence of capillary free surfaces in the absense of gravity. *Pacific J. Math.* **88** (1980), 323-361. c. 144, 162, 183.

[12] J.-T. CHEN and W. S. HUANG: Convexity of capillary surfaces in the outer space. *Invent. Math.* **67** (1982), 253-259.

[13] J.-T. CHEN: Uniqueness of minimal point and its location of capillary free surfaces over convex domains. *Astérisque* **118** (1984), 137-143.

[14] P. CONCUS: Static menisci in a vertical right circular cylinder. *J. Fluid Mech.* **34** (1968), 481-485.

[15] P. CONCUS and R. FINN: On capillary free surfaces in the absense of gravity. *Acta math.* **132** (1974), 177-198.

[16] P. Concus and R. Finn: On capillary free surfaces in a gravitational field. *Acta Math.* **132** (1974), 207-223.

[17] P. Concus and R. Finn: A singular solution of the capillary equation. I: Existence. *Invent. Math.* **29** (1975), 143-148.

[18] P. Concus and R. Finn: A singular solution of the capillary equation. II: Uniqueness. *Invent. Math.* **29** (1975), 149-160.

[19] P. Concus and R. Finn: The shape of a pendent liquid drop. *Philos. Trans. Roy. Soc. London Ser.* **A. 292** (1979), 307-340.

[20] P. Concus and R. Finn: Continuous and discontinuous disappearance of capillary surfaces. *Variational Methods for Free Surface Interface*, Springer-Verlag, 1987.

[21] P. Concus and R. Finn: Instability of certain capillary surfaces Manuscr. Math. 63 (1989) 209-213.

[22] M. Emmer: Esistenza, unicità e regolarità nelle superfici di equilibrio nei capillari. *Ann. Univ. Ferrara Sez. VII* **18** (1973), 79-94.

[23] M. Emmer and I. Tamanini: Storia della Capillarità (in preparation).

[24] R. Finn: Remarks relevant to minimal surfaces and to surfaces of prescribed mean curvature. *J. Analyse Math.* **14** (1965), 139-160.

[25] R. Finn: Some comparison properties and bounds for capillary surfaces. In *Complex Analysis and its Applications* (Russian). Moscow Math. Soc., volume dedicated to I. N. Vekua, Scientific Press, Moscow, 1978.

[26] R. Finn: The sessile liquid drop I: Symmetric Case. *Pacific J. Math.* **88** (1980), 541-587.

[27] R. Finn: Collection of articles on capillary surfaces. *Pac. J. Math.* **88** (1980).

[28] R. Finn: On the Laplace formula and the meniscus height for a capillary surface. *Z. Angew. Math. Mech.* **61** (1981), 165-173.

[29] R. Finn: Global size and shape estimates for symmetric sessile drops. *J. Reine Angew. Math.* **335** (1982), 9-36.

[30] R. Finn: Existence criteria for capillary free surfaces without gravity. *Indiana Univ. Math. J.* **32** (1983), 439-460.

[31] R. Finn: A subsidiary variational problem and existence criteria for capillary surfaces. *J. Reine Angew. Math.* **353** (1984), 196-214.

[32] R. Finn: On the pendent liquid drop. *Z. Anal. Anwend.* **4** (1985), 331-339.

[33] R. Finn: Moon surfaces, and boundary behaviour of capillary surfaces for perfect wetting and non wetting. *Proc. Lon. Math. Soc.*, 57 (1988) 542-576.

[34] R. Finn: Equilibrium Capillary Surfaces. *Grundlehren der Mathem. Wiss.* **284**, Springer-Verlag, New York, 1986. Russian Translation, with Appendix by H. C. Wente, Mir Publisher, Moscow, 1988.

[35] R. Finn: The inclination of an H-graph. «Proc. Conf. in honor of H. Lewy», Trento, 1986. Springer-Verlag, Lecture Notes 1340 (1988) 40-60.

[36] R. Finn: Nonuniqueness and uniqueness of capillary surfaces. *Manuscr. Math.* **61** (1988), 347-372.

[37] R. Finn and G. Gerhardt: The internal sphere condition and the capillary problem. *Ann. Mat. Pura Appl.* **112** (1977), 13-31.

[38] R. Finn and E. Giusti: On nonparametric surfaces of constant mean curvature. *Ann. Scuola Norm. Sup; Pisa* **4** (1977), 13-31.

[39] R. Finn and J. F. Hwang: On the comparison principle for capillary surfaces. *J. Fac. Sci. Univ. Tokyo,* 36 (1989) 131-134.

[40] Galileo Galilei: Discorso, Don Cosimo II, 1612.

[41] C. F. GAUSS: Principia Generalia Theoriae Figurae Fluidorum. *Comment. Soc. Reffiae Scient. Gottingensis Rec.* **7** (1830). Reprinted as «Grundlagen einer Theorie der Gestalt von Flüssigkeiten im Zustand des Gleichgewichtes», in Ostwald's Klassiker der exakten Wissenschaften, vol. 135. W. Engelmann, Leipzig, 1903.

[42] C. GERHARDT: Existence and regularity of capillary surfaces. *Boll. Un. Mat. Ital.* **10** (1974), 317-335.

[43] C. GERHARDT: Boundary value problems for surfaces of prescribed mean curvature. *J. Math. Pures Appl.* **58** (1979), 75-109.

[44] M. GIAQUINTA: Regolarità delle superfici $BV(\Omega)$ con curvatura media assengnata. *Boll. Un. Mat. Ital.* **8** (1973), 567-578.

[45] E. GIUSTI: Superfici cartesiane di area minima. *Rend. Sem. Mat. Fis. Milano* **40** (1970), 3-21.

[46] E. GIUSTI: Boundary value problems for non-parametric surfaces of prescribed mean curvature. *Ann. Scuola Norm. Sup. Pisa* **3** (1976), 501-548.

[47] E. GIUSTI: On the equation of surfaces of prescribed mean curvature: existence and uniqueness without boundary conditions. *Invent. Math.* **46** (1978), 111-137.

[48] E. GIUSTI: Generalized solutions to the mean curvature equation. *Pacific J. Math.* **88** (1980), 297-322.

[49] E. GIUSTI: The equilibrium configuration of liquid drops. *J. Reine Angew. Math.* **321** (1981), 53-63.

[50] E. GIUSTI: *Minimal Surfaces and Functions of Bounded Variation.* Birkhäuser, Boston, 1984.

[51] E. GONZALES: Sul problem della goccia appoggiata. *Rend. Sem. Mat. Univ. Padova* **55** (1976), 289-302.

[52] E. GONZALES, U. MASSARI and I. TAMANINI: On the regularity of the boundaries of sets minimizing perimeter with a volume constraint. *Ind. Univ. Math. J.* **32** (1983), 25-37.

[53] M. GRÜTER: Boundary reguarity for solutions of a partitioning problem. *Arch. Rat. Mech. Anal.* **97** (1987), 261-270.

[54] R. GULLIVER, S. HILDEBRANDT: Boundary configurations spanning continua of minimal surfaces. *Man. Math.* **54** (1986), 323-347.

[55] F. HAUSKSBEE: Account of the experiment on the ascent of water between two glass planes, in an hyperbolic figure. *Philos. Trans. Roy. Soc. London* **27** (1712), 539.

[56] W.-H. HUANG: Level curves and minimal points of capillary surfaces over convex domains. *Bull. Inst. Math. Acad. Sin.* **2** (1983), 390-399.

[57] N. J. KOREVAAR: On the behaviour of a capillary surface at a re-entrant corner. *Pacific J. Math.* **88** (1980), 379-385.

[58] N. J. KOREVAAR: Capillary surface continuity above irregular domains. *Comm. Partial Differential Equations* **8** (1983), 213-245.

[59] N. J. KOREVAAR: Capillary surface convexity above convex domains. *Indiana Univ. Math. J.* **32** (1983), 73-81.

[60] N. J. KOREVAAR: The normal variations technique for studying the shape of capillary surfaces. *Astérisque* **118** (1984), 189-195.

[61] N. J. KOREVAAR: An easy proof of the interior gradient bound for solutions to the prescribed mean curvature equation. *Proc. Symp. Pure Math.* **45** (1986) Part 2, 81-89.

[62] D. LANGBEIN, F. RISCHBIETER: *Form, Schwingungen und Stabilität von Flüssigkeitsgrenzflächen.* Forschungsbericht BMFT, Battelle Inst. Frankfurt/M., 1986.

[63] P. S. LAPLACE: *Traité de méchanique céleste; suppléments au Livre X, 1805 and 1806 resp. in Oeuvres Complete* Vol. 4 Gautier-Villars, Paris; see also the annoted English translation by N. Bowditch (1839); reprinted by Chelsea, New York, 1966.

[64] F. T. LIANG: *On nonparametric surfaces of constant mean curvature.* Dissertation, Stanford University, 1986; to appear.

[65] G. M. LIEBERMAN: Solvability of quasilinear elliptic equations with nonlinear boundary conditions. *Trans. Amer. Math. Soc.* **273** (1982), 753-765.

[66] G. M. LIEBERMAN: Boundary behaviour of capillary surfaces via the maximum principle. *Variational Methods for Free Surface Interfaces,* Springer-Verlag, 1987.

[67] G. M. LIEBERMAN: Gradient estimates for capillary-type problems via the maximum principle (to appear).

[68] G. M. LIEBERMAN: Hölder continuity of the gradient at a corner for the capillary problem and related results. *Pac. J. Math.* **133** (1988), 115-135.

[69] TH. LOHNSTEIN: Zur Theorie des Abtropfens mit besonderer Rücksicht auf die Bestimmung der Kapillaritätskonstanten durch Tropfenversuche. *Ann. Physik* **20** (1906), 237-268.

[70] U. MASSARI: Esistenza e regolarità delle ipersuperfici di curvatura media assegnata in R^n. *Arch. Rational Mech. Anal.* **55** (1974), 257-382.

[71] U. MASSARI, M. MIRANDA: *Minimal Surfaces of Codimension One.* North Holland Mathematics Studies 91, Elsevier Science Publ., Amsterdam, 1984.

[72] J. C. MELROSE: Model calculations for capillary condensation. *A. I. Ch. E. Journal,* (1966), 986-994.

[73] E. MIERSEMANN: On capillary free surfaces without gravity. *Z. Anal. Anwend.* **4** (1985) 429-436.

[74] E. MIERSEMANN: Asymptotic expansion in a corner for the capillary problem. *Pac. J. Math.,* **134** (1988), 299-312.

[75] E. MIERSEMANN: On the behaviour of capillaries in a corner. *Pac. J. Math.,* in press.

[76] M. MIRANDA: Analiticità delle superfici di area minima in R^n. *Rend. Acc. Naz. Lincei* **38** (1965), 632-638.

[77] M. MIRANDA: Superfici cartesiane generalizzate ed insiemi di perimetro localmente finito sui prodotti cartesiani. *Ann. Scuola Norm. Sup. Pisa* **3** (1976), 501-548.

[78] M. MIRANDA: Superfici minime illimitate. *Ann. Scuola Norm. Sup. Pisa* **4** (1977), 313-322.

[79] M. MIRANDA: A Mathematical Description of Equilibrium Surfaces. *Variational Methods of Free Surface Interfaces,* Springer-Verlag, 1987.

[80] A. D. MYSHKIS, V. G. BABSKY, N. D. KOPACHEVSKY, L. A. SLOBOZHANIN and A. D. TYUPTSOV: *Low Gravity Hydromechanics* (Russian). Scientific Press, Moscow, 1976. (English translation: Springer-Verlag, 1988.)

[81] J. W. S. RAYLEIGH: *The Theory of Sound,* Vol. 2, Dover. Publ., N. Y., 1945. See esp. §357.

[82] J. A. SEGNER: De figuris superficierum fluidarum. *Commentarii Societ. Regiae Scientiarum Göttingensis* **1** (1752), 301-372.

[83] J. B. SERRIN: A symmetry problem in potential theory. *Arch. Rational Mech. Anal.* **43** (1971), 304-318.

[84] J. B. SERRIN: The Dirichlet problem for surfaces of constant mean curvature. *Proc. Lon. Math. Soc.* (3) **21** (1970), 361-384.

[85] D. SIEGEL: Height estimates for capillary surfaces. *Pacific J. Math.* **88** (1980), 471-516.

[86] D. SIEGEL: The Behaviour of a Capillary Surface for Small Bond Number. *Variational Methods for Free Surface Interface*, Springer-Verlag, 1987.

[87] D. SIEGEL: *Explicit estimates of a symmetric capillary surface for small Bond number.* Preprint, Univer. of Waterloo, to appear.

[88] L. SIMON: Regularity of capillary surfaces over domains with corners. *Pacific J. Math.* **88** (1980), 363-377.

[89] J. SPRUCK: On the existence of a capillary surface with a prescribed angle of contact. *Comm. Pure Appl. Math.* **28** (1975), 189-200.

[90] L.-F. TAM: The behaviour of capillary surfaces as gravity tends to zero. *Comm. P.D.E.* **11** (1986), 851-901.

[91] L.-F. TAM: Regularity of capillary surfaces over domains with corners. *Pac. J. Math.* **124** (1986), 469-482.

[92] L.-F. TAM: On the Uniqueness of Capillary Surfaces. *Variational Methods for Free Surface Interface*, Springer-Verlag, 1987.

[93] I. TAMANINI: Interfaces of Prescribed Mean Curvature. *Variational Methods for Free Surface Interface*, Springer-Verlag, 1987.

[94] I. TAMANINI: *Variational problems of least area with constraints.* Preprint, Univ. di Trento, to appear.

[95] B. TAYLOR: Concerning the ascent of water between two glass planes. *Philos. Trans. Roy. Soc. London* **27** (1712), 538.

[96] J. E. TAYLOR: Boundary regularity for solutions to various capillarity and free boundary problems. *Comm. Partial Differential Equations* **2** (1977), 323-357.

[97] N. N. URAL'TSEVA: Solution of the capillary problem (Russian). *Vestnik Leningrad Univ* **19** (1973), 54-64.

[98] T. I. VOGEL: Symmetric unbounded liquid bridges. *Pac. J. Math.* **103** (1982), 205-241.

[99] T. I. VOGEL: Asymmetric unbounded liquid bridges. *Ann. Scuola Norm. Sup. Pisa IV* **9** (1982), 433-442.

[100] T. I. VOGEL: Unbounded parametric surfaces of prescribed mean curvature. *Indiana Univ. Math.* J. **31** (1982), 281-288.

[101] T. I. VOGEL: Uniqueness for certain surfaces of prescribed mean curvature. *Pac. J. Math.* **134** (1988), 197-207.

[102] T. I. VOGEL: Stability of a liquid drop trapped between two parallel planes. *SIAM J. Appl. Math.* **47** (1987), 516-525.

[103] T. I. VOGEL: *Stability of a liquid drop trapped between two parallel planes II: general contact angles.* Preprint, Texas A & M Univ., to appear.

[104] H. C. WENTE: An existence theorem for surfaces in equilibrium satisfying a volume constraint. *Arch. Rational Mech. Anal.* **50** (1973), 139-158.

[105] H. C. WENTE: The symmetry of sessile and pendent drops. *Pacific J. Math.* **88** (1980), 387-397.

[106] H. C. WENTE: The stability of the axially symmetric pendent drop. *Pacific J. Math.* **88** (1980), 421-470.

[107] T. YOUNG: An essay on the cohesion of fluids. *Philos. Trans. Roy. Soc. London* **95** (1805), 65-87.

ON THE SIMPLE SHAPE OF STABLE EQUILIBRIA

BERNHARD KAWOHL

Abstract. *A standard argument in the theory of symmetrization goes as follows. Let* u *be a global minimizer of a variational problem, i.e. let* u *minimize a functional* $J : A \to \mathbf{R}$, *and suppose that* u^*, *the symmetrization of* u *is an admissible function for the variational problem, i.e.* $u^* \in A$. *Then (cum grano salis)* $u = u^*$. *Similar results are known for local minima of some variational problems, but the standard symmetrization arguments fail. It is therefore desirable to consider a homotopy between* u *and* u^* *in the class of functions which are equimeasurable to* u. *I address this question and give some partial answers to it, which have been obtained in joint work with* H. *Matano. These answers allow to prove symmetry and other properties for local minimizers of variational problems with many side constraints. The talk contains also a little survey of related results and problems.*

Consider the partial differential equation

$$u_t - \Delta u + W'(u) = 0 \quad \text{in } \Omega \times \mathbf{R}^+ \tag{1}$$

where $\Omega \subset \mathbf{R}^n$ is a bounded domain and W' is locally Lipschitz continuous, under boundary conditions of Dirichlet type

$$u = 0 \quad \text{on } \partial\Omega \times \mathbf{R}^+, \tag{2}$$

or Neumann type

$$\frac{\partial u}{\partial n} = 0 \quad \text{on } \partial\Omega \times \mathbf{R}^+. \tag{3}$$

Any linearly stable equilibrium solution of (1) (2) or (1) (3) will satisfy the elliptic equation

$$-\Delta u + W'(u) = 0 \quad \text{in } \Omega \tag{4}$$

as well as (2) or (3) and will be a local minimizer of the functional

$$J(v) = \int_\Omega \left\{ \frac{1}{2}|\nabla v|^2 + W(v) \right\} dx \tag{5}$$

over $W_0^{1,2}(\Omega)$ or $W^{1,2}(\Omega)$, i.e. $J(u) \leq J(v)$ for any v in a sufficiently small neighborhood of u.

It is known that solutions of (4) can in general have many «oscillations». One can easily see this in the special case $u'' + \lambda u = 0$ on $(0, \pi)$. Stable equilibria, however, have «simple» shapes, and they are the subject of my lecture. At the end of this talk I shall also present two innocent looking but challenging open problems on shapes for the case $f(u) \equiv 0$.

Let me first collect some known results on special cases of (4).

THEOREM 1. *Let u be a linearly stable solution of (4).*

a1) If u satisfies (2) and $n = 1$; then u is of only one sign, say $u \geq 0$, and u is quasiconcave, i.e. u' changes sign at most once, or, equivalently, the sets $\{x \in \Omega / u(x) > c\}$ are convex.

a2) If u satisfies (2) and $n = 2$, and if Ω is convex and $u > 0$ in Ω, then $\nabla u = 0$ in only one point.

b) If u satisfies (3) and Ω is convex, then u is constant for any n

Result a1) is due to Maginu [Ma], a2) was derived by H. Matano [M3]. Matano has also found stable solutions of alternate sign. Result b) was first established for $n = 1$ by N. Chafee [C] and later extended to general n by H. Matano [M1] and Casten and Holland [CH] independently. Actually the convexity assumption on Ω in b) can be weakened to pseudoconvexity (the mean curvature of $\partial\Omega$ is nonnegative), but this would make is slightly harder to see the analogy between a2) and b).

From now on I concentrate on the variational problem (5) rather than on its Euler equation (4). There are a number of results on local minimizers of variational problems under side constraints, e.g. under

$$\int_\Omega u \, dx = 0 \tag{6}$$

These results suggest that one should expect simple shapes for constrained problems as well.

PROPOSITION 2. *Let u be a local minimizer of (5) under side constraint (6).*

a) If u satisfies (2), as to my knowledge, nothing seems to be previously known about the general shape of u.

b1) If u satisfies (3) and $n = 1$, then u is monotone, i.e. u' does not change sign.

b2) If u satisfies (3) and $n = 2$, and if Ω has certain geometric properties (e.g. Ω is a disc or a rectangle), then u is monotone in some directions, i.e. some partial derivatives of u do not change sign.

Statement b1) can be found in papers on the Cahn-Hilliard theory of phase transitions, see V. Alexiades and C. E. Aifantes [AA] or J. Carr, M. E. Gurtin and M. Slemrod [CGS].

Statement b2) is due to M. E. Gurtin and H. Matano [GM].

When H. Matano told me about b2) in 1984 we decided to work on the shape of stable equilibria with rearrangement methods. The following main result was obtained in collaboration with H. Matano. I regret that he was unable to attend this meeting.

THEOREM 3. *Let $n = 1; \Omega = (a,b), W \in C^{1,1}(\mathbf{R},\mathbf{R})$, $g_i \in C^2(\mathbf{R},\mathbf{R})$ and $c_i \in \mathbf{R}, i = 1,\ldots,N; N \in \mathbb{N}$. Suppose that*

$$\{g_i'\}_{i=1,\ldots,N} \quad \text{are linearly independent functions} \tag{7}$$

and set

$$G = \bigcap_{i=1}^{N} \left\{ \int_a^b g_i(u) = c_i \right\}. \tag{8}$$

a) If u is a local minimizer of (5) on $W_0^{1,2}(\Omega) \cap G$, then the sets $\{x \in (a,b) | u(x) > 0\}$ and $\{x \in (a,b) | u(x) < 0\}$ are simply connected and u is symmetrically decreasing resp. increasing on either one of them.

b) If u is a local minimizer of (5) on $W^{1,2}(\Omega) \cap G$, then u is monotone in (a,b).

REMARK 4. Notice that Proposition 2b 1) is contained in Theorem 3.

BEGINNING OF PROOF OF THEOREM 3. Assumption (7) makes sure that the manifold G is a transversal intersection of the manifolds

$$G_i = \left\{ \int_a^b g_i(u) = c_i \right\}, i = 1,\ldots,N$$

in function space. Under this assumption, u has to satisfy the Euler-Lagrange equation

$$-u'' + W'(u) + \sum_{i=1}^{n} \lambda_i g_i'(u) = 0 \quad \text{in } (a,b). \tag{9}$$

Suppose that u' vanishes in more than finitely many points. Then there exists $\{x_n\}_{n\in N} \subset (a,b)$ converging to x_∞ such that $u'(x_n) = 0 = u'(x_\infty)$. Consequently there exists $\{y_n\}_{n\in N} \subset (a,b)$ converging to x_∞ such that $u''(y_n) = 0 = u''(x_\infty)$. But now

$$f(u(x)) = W'(u(x)) + \sum_{i=1}^{n} \lambda_i g_i'(u(x)) = 0$$

vanishes at x_∞. Set $w(x) = u'(x)$.

Then $w' = f\left(\int_a^x w - u(a)\right)$ and $w(x_\infty) = w'(x_\infty) = 0$. Now the uniqueness theorem for ordinary differential equation implies $w \equiv 0$, i.e. u is constant. Therefore either u is constant, in which case there is nothing more to prove or

(10) u' vanishes finitely often in (a,b).

REMARK 5. At this point let me pause with the proof for a while and remark that a weaker statement than Theorem 3 can now be proved: If u is a global (not local) minimizer of (5) on $W^{1,2}(\Omega) \cap G$, then u is monotone in (a,b). This follows from a rearrangement result. Let u^* be the monotone decreasing rearrangement of u (see [K] for definitions). Then $u \in G$ implies $u^* \in G$, and if u is not monotone , $J(u^*)$ must be strictly less than $J(u)$. This follows from [K, p. 34] and the contradicts the fact that u is a global minimizer. Notice that u is «smooth» in the sense required in [K] because of property (10).

Similarly one can conclude that global minimizers u of (5) on $W_0^{1,2} \cap G$ are symmetrically decreasing (resp. increasing) on each component of $\{u > 0\}$ (resp. $\{u < 0\}$), see [K, p. 39], and that $\{u > 0\}$ (resp. $\{u < 0\}$) has only component. Therefore Theorem 3 is certainly true for global minimizers.

The argument in Remark 5 has no obvious extension to local minimizers . In fact, in a situation like the one depicted in Fig. 1 below is conceivable.

Here u^* is a rearrangement of u with the properties described in Theorem 3a) resp. 3b), and u is in $W_0^{1,2}(\Omega) \cap G$ resp. $W^{1,2}(\Omega) \cap G$. Suppose u is local minimizer but violates the statements of Theorem 3, i.e.

$$\int_\Omega |\nabla u|^2 \mathrm{d}x > \int_\Omega |\nabla u^*|^2 \mathrm{d}x.$$

Can one construct a continuous path $\{u_t\}_{0\le t\le 1}$ in $W_0^{1,2}(\Omega)$ resp. $W^{1,2}(\Omega)$ such that the following properties i) - v) hold true?

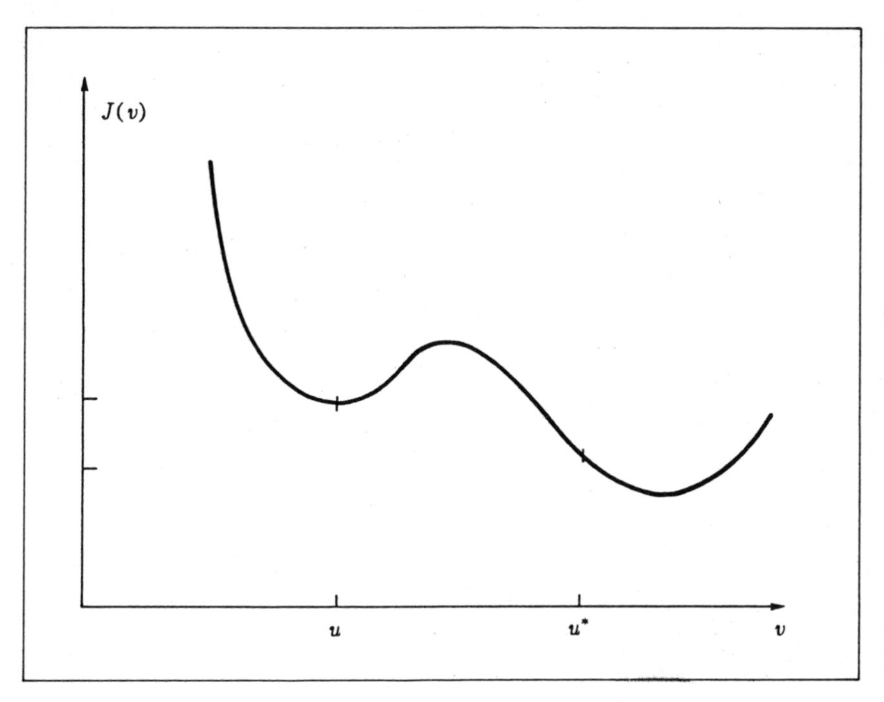

Fig. 1.

i) $u_0(x) = u(x)$;

ii) $u_1(x) = u^*(x)$;

iii) u and u_t are equimeasurable for $t \in [0,1]$;

iv) $D(t) = \int_\Omega \left|\frac{du_t}{dx}\right|^2 dx$ is monotone noncreasing in $t \in [0,1]$;

v) $D(t) < D(0)$ for every sufficiently small $t < 0$.

Then one would have a prooof of Theorem 3 for local minimizers, because $u_t \in G$ and $J(u_t) < J(u)$ for some small t.

H. Matano and I have been able to construct such a path and thus to complete the proof of Theorem 3. The construction is very technical and delicate, and therefore I shall only present some heuristic argument in this lecture. At the same time, I must remind you that heuristic arguments are not proofs and can be misleading; see e.g. [BZ] for pointing out a small but essential oversight in a related theorem. The proof and analytical description of the construction below is contained in [KM].

So we want to «continuously symmetrize» a function. An established way to rearrange functions is via rearrangements of level sets. There are attemps in the literature, to continuosly rearrange level sets. Polya and Szegö [PS, Note B] and McNabb [Mc] have been able to construct a path which satisfies i)- iv), however

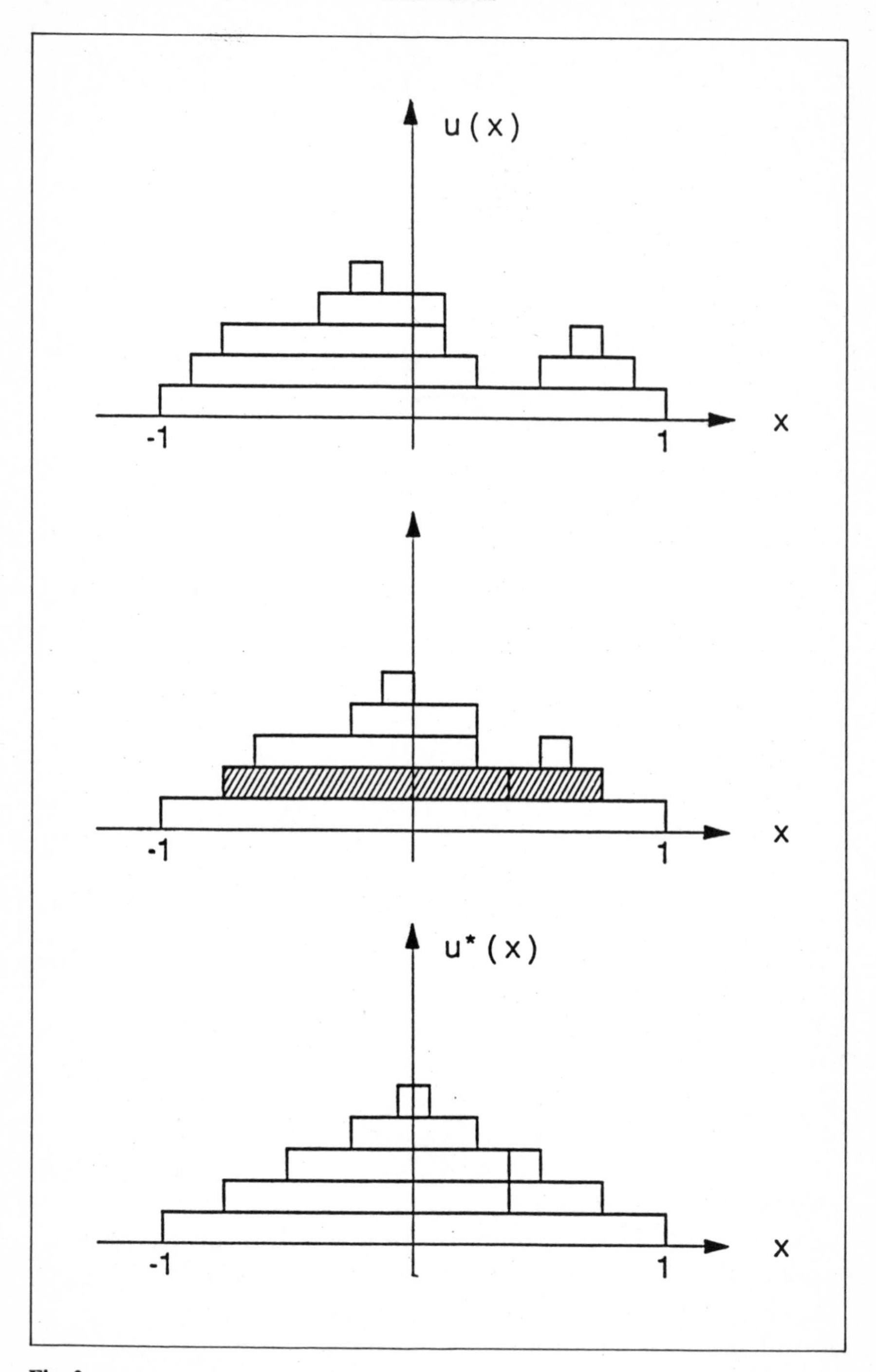

Fig. 2.

only if u has convex level sets. The approach for quasiconcave functions u is not suitable for our purposes. Therefore, and to earn my honorarium, I shall continue with an explanation of our own construction.

For simplicity of the exposition set $(a,b) = (-1,1)$ and symmetrize a step function layer by layer, as in Fig. 2, starting at the bottom. Incidentally, one could symmetrize the skyline of Manhattan in this fashion, but the effect would be somewhat disappointing.

On an infinitesimal scale, this would be a continuous symmetrization.

In Fig. 3 it is shown how a nonnegative, piecewise linear, continuous function with symmetric support is continuously symmetrized «from bottom to top». If the graph has portions with horizontal shape, this naive algorithm must be modified to make it a continuous one. In that case parts of the graph are horizontally slided in a continuous fashion without altering the Dirichlet integral. Again I have to refer to [KM] for details.

So in principle we understand how to continuously symmetrize a function in such a way that properties i)-iv) hold. Property v) will be dealt with later when we use additional information (Proposition 2) on the regularity of u. Whithout such extra information we cannot expect v) to hold. This is illustrated by the following counterexample, whose graphs are depicted in Fig. 4 below.

Let $\Omega = (-1,1)$ and let $Q \cap (-1,0) = \{a_m | m \in N\}$ be the set of all rational numbers in $(-1,0)$. Let $\epsilon_1, \epsilon_2, \epsilon_3, \ldots$ be positive numbers with $\sum_{m=1}^{\infty} 2\epsilon_m < 1$ and define $A_1 = (-1,0) \cap \bigcup_{m \in N} (a_m - \epsilon_m, a_m + \epsilon_m)$. Then $0 < \mu(A_1) < 1$, and A_1 is a dense open subset of $(-1,0)$. Next set $A_2 = \{-1 + \lambda(x+1) | x \in A_1\}$; where $\lambda = 2/(1 + \lambda(A_1)) > 1$. It is clear that A_2 is a dense open subset of $(-1, \lambda - 1)$ and that $\mu(A_2) = \mu([\lambda - 1, 1])$.

Now define

$$u(x) = \begin{cases} \mu(A_2 \cap [-1,x]) & \text{if } x \leq \lambda - 1, \\ \mu([x,1]) & \text{if } \lambda - 1 < x \leq 1. \end{cases} \tag{11}$$

It is easily seen that $u \in X_0$ and that u is strictly increasing in $[-1, \lambda - 1]$ and strictly decreasing in $[\lambda - 1, 1]$. Therefore the set $\{u = c\}$ contains at most two points for every $c \leq 0$. It is also clear that $\left\{ \frac{du}{dx} = 0 \right\}$ is a nowhere dense subset of Ω (i.e. its closure has an empty interior). Moreover a direct calculation shows that $\int_\Omega \left(\frac{du^*}{dx} \right)^2 dx = \int_\Omega \left(\frac{du}{dx} \right)^2 dx$. Nonetheless we have $u \neq u^*$ since $u(x) \not\equiv u(-x)$.

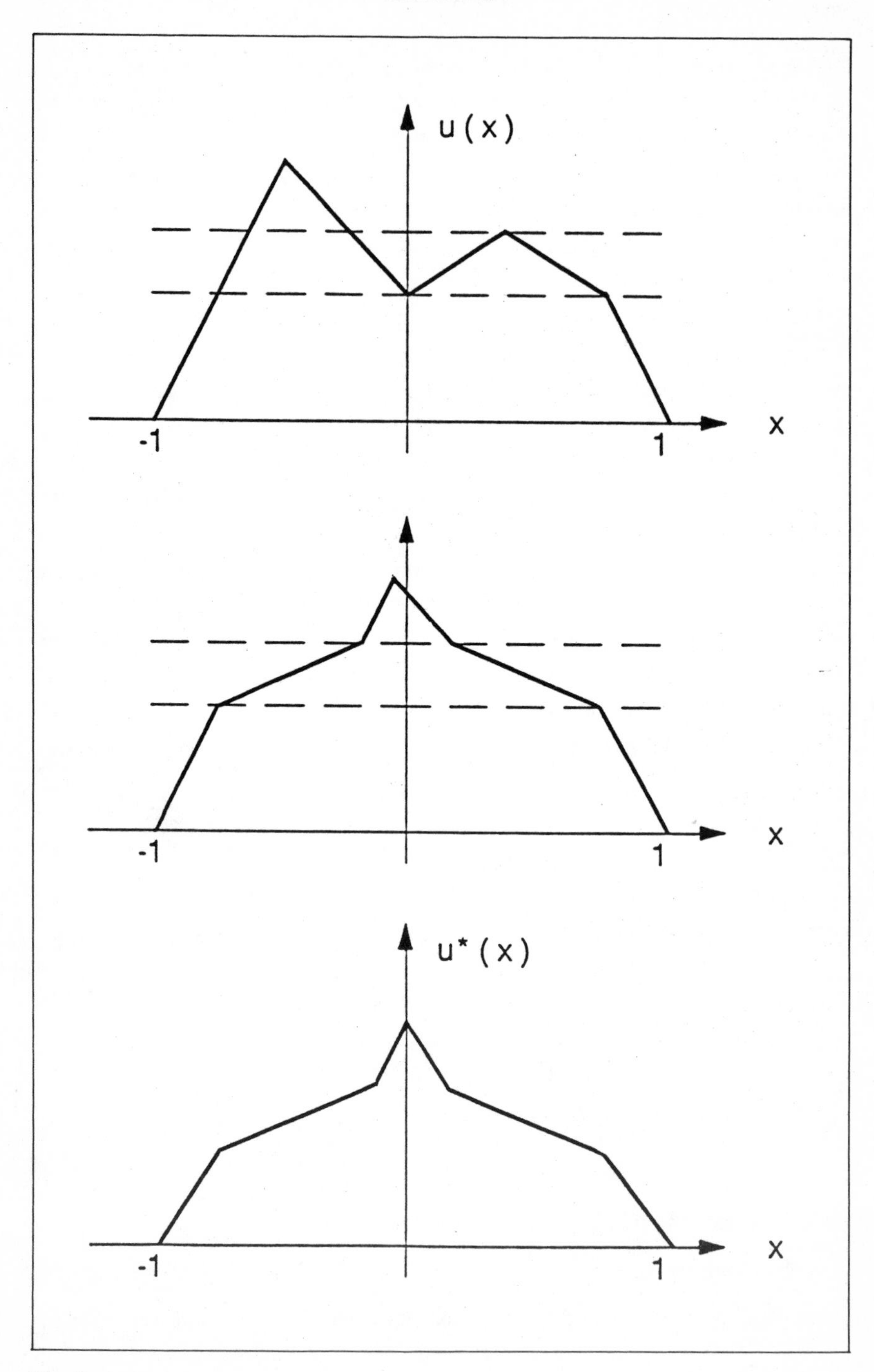

Fig. 3.

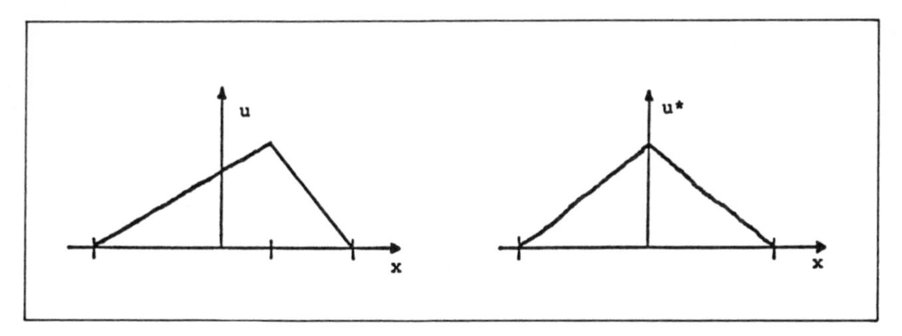

Fig. 4.

Notice that

$$u' = \begin{cases} 1 & \text{if } x \in A_2 \qquad \text{(dense open)} \\ 0 & \text{else} \qquad \text{(nowhere dense)} \end{cases} \text{ and } x < \lambda - 1$$

for this counterexample.

Next I attempt to illustrate in Fig. 5, how one can rearrange a function with nonempty components $\{u < 0\}$ and $\{u > 0\}$ into one that satisfies statement a) of Theorem 3.

In passing from Fig. 5 a) to Fig. 5 b) I have shifted the bold part of the graph of u to the northeast and reflected it across the w-axis. Then I have symmetrized the graph w from height zero up to the point A, and in this way I have generated the function w_T in Fig. 5 c). In Fig. 5d) the (rearranged) bold part of w_T is shifted southeastward, back into place. Moreover the remaining part of the graph u is added to the right of F. Now one can symmetrize the negative part of v_T about the center of its support and arrive at Fig. 5e).

The previous heuristic arguments should have convinced you that one can indeed construct a continuous path which connects u to rearrangement u^{**} in such a way that properties i)-iv) hold. It remains to show v). So suppose that u is a local minimizer of (5) on $W_0^{1,2}(\Omega) \cap G$, but fails to satisfy the claims of Theorem 3 a). Then u has more than 2 local extrema. Now a careful analysis shows that one can rearrange u «locally near the height of one of the extrema» in such a way that property v) holds. Afterwards one can continue with the above construction for arbitrary $W_0^{1,2}(\Omega)$-functions. It is essential for this analysis, that pathologies like the ones depicted in Fig. 4 cannot occur. Fortunately we known from Proposition 2 that u cannot have too many critical points. This completes a sketch of proof for Theorem 3 a).

Theorem 3 b) is derived in a similar manner. Again a picture says more than thousand words. Suppose that $u(x)$ is not a monotone function and that a global maximum of u lies to the left (resp. right) of a global minimum. Then I want to

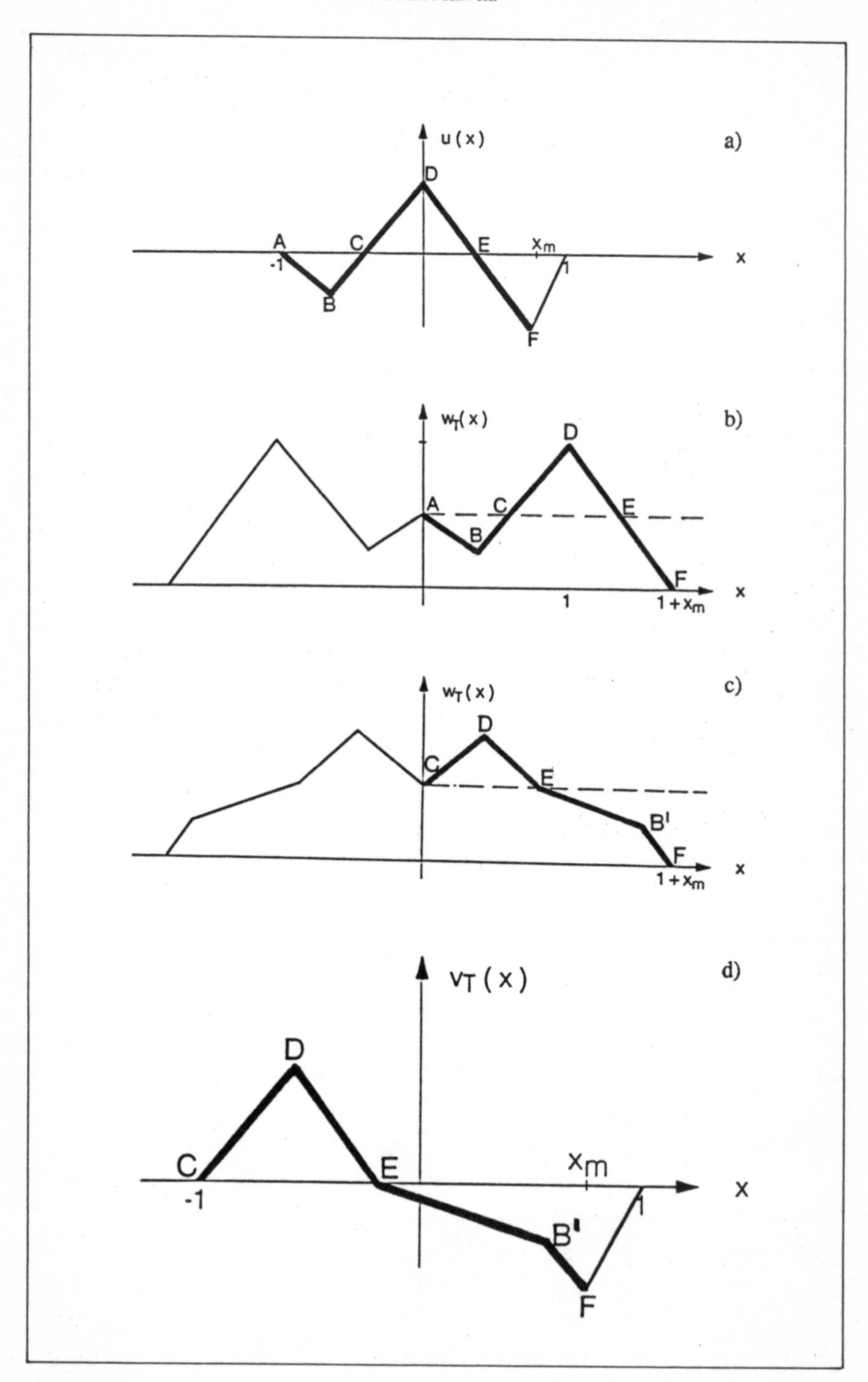
u(x)
a)
D
A
C
E
x_m
x
-1
1
B
F
$w_T(x)$
b)
D
A
C
E
B
F
x
1
1+x_m
$w_T(x)$
c)
D
C
E
B'
F
x
1+x_m
$v_T(x)$
d)
D
C
E
x_m
x
-1
1
B'
F

Fig. 5.

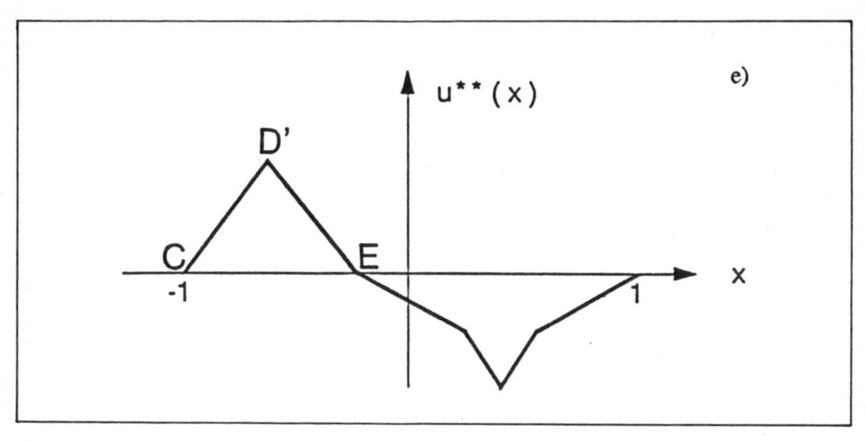

Fig. 5. (segue).

continuously deform u into its monotone decreasing (resp. increasing) rearrangement. I reflect u across its left (resp. right) endpoint and connect the symmetrically extended function w (see Fig. 6 b) to its symmetrically decreasing rearrangement. In a first stage I rearrange the bold part of graph of u in Fig. 6 b) and arrive at Fig. 6 c). Then I turn the graph of v_T upside down, reflect it across C and slide it upwards to obtain a function $\tilde{w}$ in Fig. 6 d). A continuous symmetrization of $\tilde{w}$ and another reflection of the graph allows me to arrive in Fig. 6 e) at u_*, the monotone decreasing rearrangement of u. An analytical description of these graphic ideas can be found in [KM] and provides a proof of Theorem 3 b) as well.

Let me now return to parabolic problems again and present two open problems with partial answers.

OPEN PROBLEM 1. Consider the linear heat equation

$$u_t - \Delta u = 0 \qquad \text{in } \Omega \times \mathbf{R}^+ \tag{12}$$

under Dirichlet boundary conditions

$$u = 0 \qquad \text{on } \partial\Omega \times \mathbf{R}^+ \tag{2}$$

and with initial data

$$u(x,0) = u_0(x) \qquad \text{in } \Omega \tag{13}$$

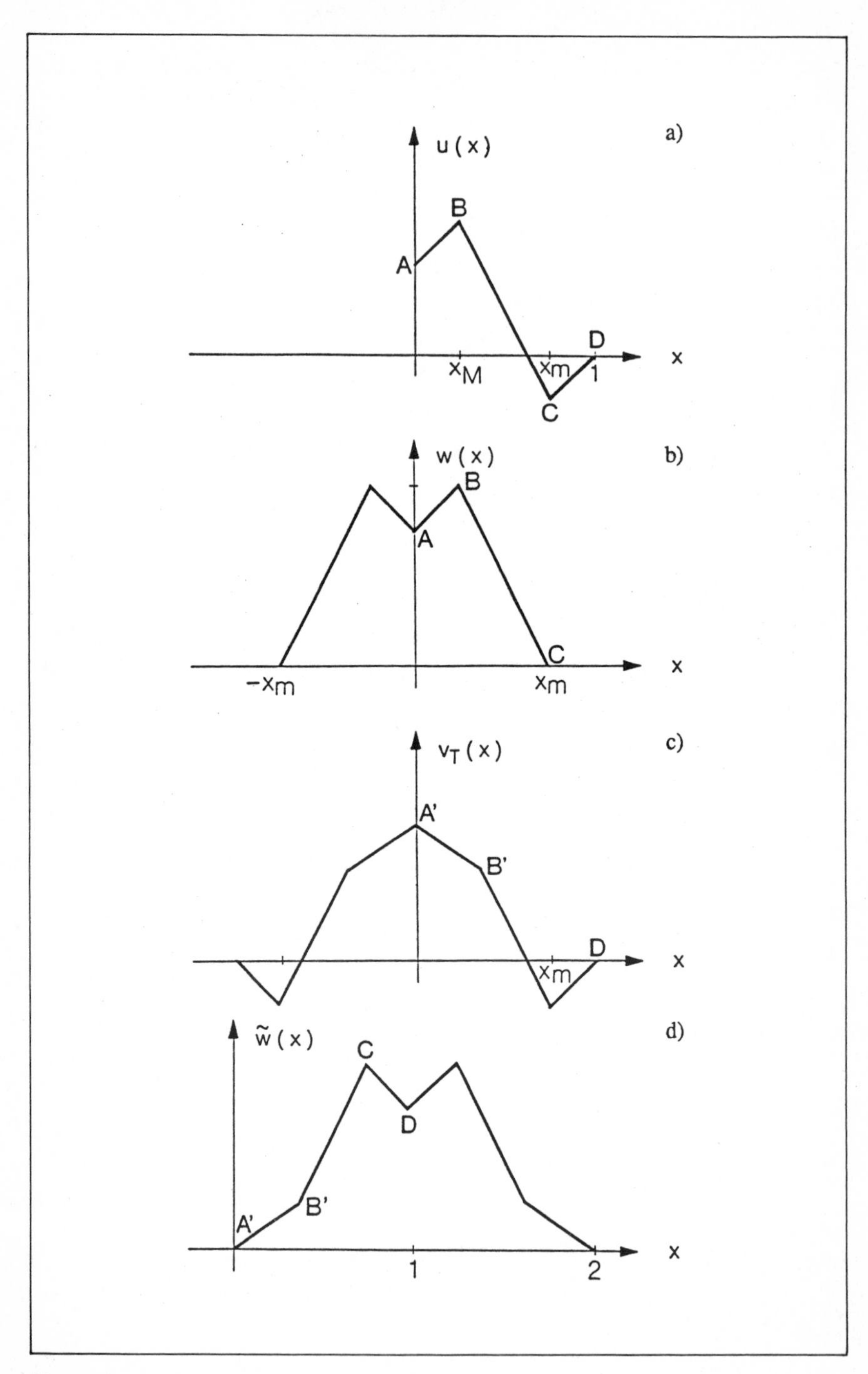

Fig. 6.

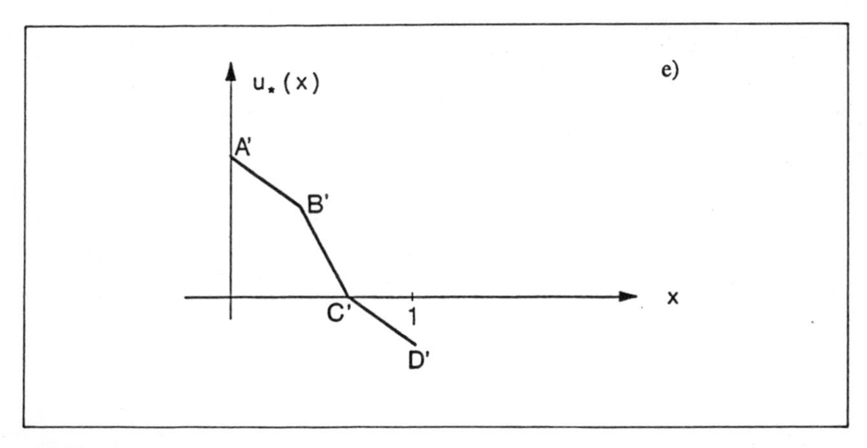

Fig. 6. (segue).

Suppose that $\Omega \subset \mathbf{R}^n$ is a convex bounded domain and that $u_0(\cdot)$ is a quasiconcave function of x, i.e. the sets $\{x \in \Omega \,|\, u_0(x) \geq C\}$ are convex. Suppose moreover that $u_0 \geq 0$ in Ω and $u_0 = 0$ on $\partial\Omega$.

Is it true that $u(\cdot, t)$ is quasiconcave in x for any $t > 0$?

This question has puzzled quite a few people, and for $n = 1$ the anwer is positive, see [S, P, N, M2, Ke, A]. For $n = 2$ the problem appears to be open. A related result, however, is the following one.

If $u_0(x)$ is logconcave in x, then $u(\cdot, t)$ is logconcave in x for all $t > 0$.

For proofs I refer to Brascamp and Lieb [BL] and to Korevaar [Ko].

OPEN PROBLEM 2. Consider the linear heat equation (12) under Neumann boundary conditions

$$\frac{\partial u}{\partial n} = 0 \qquad \text{on } \partial\Omega \times \mathbf{R}^+ \tag{3}$$

and with initial data $u_0(x)$. For every $t \geq 0$ let $x(t)$ denote the «hot spot», i.e. the point in which $u(\cdot, t)$ attains its global maximum.

In numerical calculations and in physical experiments the hot spot moves to the boundary and settles on the boundary as t goes to infinity. For a possible explanation of this behaviour one can look at the Fourier series representation

$$u(x,t) = \sum_{j=1}^{\infty} \alpha_j e^{-\nu_j t} u_j(x), \tag{14}$$

where α_j are the Fourier coefficients of $u_0(x)$; ν_j the eigenvalues and u_j the eigenvalues of the Laplace operator under Neumann boundary conditions. Recall that

$\nu_1 = 0$ and $u_1 = \text{const}$., so that the first term in the expansion (14) does not contribute anything to the asymptotic shape of u.

If $\alpha_2 \neq 0$, however, the shape of the second eigenfunction will dominate the asymptotic shape of u, since all the other terms decay faster than $e^{-\nu_2 t}$. Thus one is lead to a conjecture of J. Rauch (see [K]):

Is it true that u_2, the second eigenfunction of the Laplace operator under Neumann boundary conditions, attains its maximum on the boundary?

There are positive answers to this problem if $n = 1$ or if $\Omega \subset \mathbb{R}^n$ with $n \geq 2$ and Ω is cylindrical, but for general domains Ω the conjecture appears to be open.

REFERENCES

[A] S. ANGENENT: The zeroset of a solution of a parabolic equation. *J. reine angew. Math.* **390** (1988), 79-86.

[AA] V. ALEXIADES and E. C. AIFANTIS: On the thermodynamic theory of fluid interfaces equilibrium solutions, and minimizers. *J. of Colloid and Interface Science* **111** (1986), 119-132.

[BL] H. J. BRASCAMP and E. H. LIEB: On extensions of the Brunn-Minkowski and Prékopa-Leindler theorems, including inequalities for log-concave functions, and with an application to the diffusion equation. *J. Funct. Anal.* **22** (1976), 366-389.

[BZ] J. E. BROTHERS and W. P. ZIEMER: Minimal rearrangements of Sobolev functions. *J. reine angew. Math.* **384** (1988), 153-179.

[CGS] J. CARR, M. E. GURTIN and M. SLEMROD: Structured phase transitions on a finite interval. *Arch. Ration. Mech. Anal.* **86** (1984), 317-351.

[CH] R. G. CASTEN and C. J. HOLLAND: Instability results for reaction diffusion equations with Neumann boundary conditions. *J. Differ. Equations* **27** (1978), 266-273.

[C] N. CHAFEE: Asymptotic behaviour for solutions of a one-dimensional parabolic equation with homogeneous Neumann boundary conditions. *J. Differ. Equations* **18** (1975), 111-135.

[FM] A. FRIEDMAN and B. McLEOD: Strict inequalities for integrals of decreasingly rearranged functions. *Proc. Roy. Soc. Edinburgh* **A 102** (1986), 227-289.

[GM] M. E. GURTIN and H. MATANO: On the structure of equilibrium phase transitions within the gradient theory of fluids. *Quarterly of Applied Math.* **46** (1988), 301-317.

[K] B. KAWOHL: Rearrangements and convexity of level sets in PDE. Springer Lecture Notes in Math. 1150 (1985).

[KM] B. KAWOHL and H. MATANO: Continuous equimeasurable rearrangement and applications to variational problems (in preparation).

[Ke] G. KEADY: The persistence of log-concavity for positive solutions of the one-dimensional equation. Manuscript 1987, submitted to *J. Austr. Math. Soc. A.*

[Ko] N. KOREVAAR: Convex solutions to nonlinear elliptic and parabolic boundary value problems. *Indiana Univ. Math. J.* **32** (1983), 606-614.

[Ma] K. MAGINU: Stability of stationary solutions of a semilinear parabolic differential equation. *Math. Anal. Appl.* **63** (1978), 224-243.

[M1] H. MATANO: Asymptotic behaviour and stability of solutions of semilinear diffusion equations. *Publ. Res. Inst. Math. Sci.* **15** (1979), 401-454.

[M2] H. MATANO: Nonincrease of the lap number of a solution for a one-dimensional semilinear parabolic equation. *J. Fac. Sci. Univ. of Tokyo Sec.* **IA 29** (1982), 401-441.

[M3] H. MATANO: personal communication, 1987.

[Mc] A. MCNABB: Partial Steiner symmetrization and some conduction problems. *J. Math. Anal. Appl.* **17** (1967), 221-227.

[N] K. NICKEL: Gestaltaussagen über Lösungen parabolischer Differentialgleichungen. *J. reine angew. Math.* **211** (1962), 78-94.

[P] G. POLYA: Qualitatives über den Wärmeausgleich. *Zeitschr. Angew. Math. Mech.* **10** (1930), 353-360.

[PS] G. POLYA and G. SZEGÖ: Isoperimetric inequalities in mathematical physics. *Ann. Math. Studies* **27** (1952), Princepton Univ. Press.

[S] C. STURM: Sur une classe d'equations à differences partielles. *J. Math. Pures Appl.* **1** (1836), 373-444.

THE (NON) CONTINUITY OF SYMMETRIC DECREASING REARRANGEMENT

FREDERICK J. ALMGREN JR. - ELLIOTT H. LIEB

Abstract. *The operation $\mathcal{R}$ of symmetric decreasing rearrangement maps $W^{1,p}(\mathbf{R}^n)$ to $W^{1,p}(\mathbf{R}^n)$. Even though it is norm decreasing we show that $\mathcal{R}$ is not continuous for $n \geq 2$. The functions at which $\mathcal{R}$ is continuous are precisely characterized by a new property called* **co-area regularity**. *Every sufficiently differentiable function is co-area regular, and both the regular and the irregular functions are dense in $W^{1,p}(\mathbf{R}^n)$.*

1. INTRODUCTION

Suppose $f(x^1, x^2) \geq 0$ is a continuously differentiable function supported in the unit disk in the plane. Its rearrangement is the rotationally invariant function $f^*(x^1, x^2)$ whose level sets are circles enclosing the same area as the level sets of f, i.e.

$$\mathcal{L}^2\{(x^1, x^2) : f(x^1, x^2) > y\} = \mathcal{L}^2\{(x^1, x^2) : f^*(x^1, x^2) > y\}$$

for each positive height y ($\mathcal{L}^n$ denotes Lebsgue over $\mathbf{R}^n$). Such rearrangement preserves L^p norms, i.e.

$$\int |f^*|^p d\,\mathcal{L}^2 = \int |f|^p d\,\mathcal{L}^2 \quad (1 \leq p < \infty)$$

but decreases convex gradient integrals, e.g.

$$\int |\nabla f^*|^p d\,\mathcal{L}^2 \leq \int |\nabla f|^p d\,\mathcal{L}^2 .$$

Now suppose that $f_j(x^1, x^2) \geq 0$ $(j = 1, 2, 3, \ldots)$ is a sequence of continuosly differentiable functions also supported in the unit disk which converge uniformly together with first derivatives to f, i.e.

$$f_j(x^1, x^2) \to f(x^1, x^2) \quad \text{and} \quad \nabla f_j(x^1, x^2) \to \nabla f(x^1, x^2)$$

uniformly in (x^1, x^2) as $j \to \infty$. It is not difficult to check that the symmetrized functions also converge uniformly. The real question is about convergence of the derivatives of the symmetrized functions. It is certainly plausible that they should converge strongly (we believed it for some time). Our principal new result is that *the derivatives of the symmetrized functions need not converge strongly,* e.g. for special f's and f_j's satisfying our conditions above it can happen that for every p

$$\liminf_{j\to\infty} \int |\nabla f_j^* - \nabla f^*|^p d\mathcal{L}^2 > 0 .$$

Furthermore, we are able to characterize exactly those f's for which convergence is assured and for which it can fail.

The general notion of the symmetric decreasing rearrangement f^* of a function $f : \mathbf{R}^n \to \mathbf{R}^+$ is important in various parts of analysis. For example, various rotationally invariant variational integrals (like the gradient norms mentioned above) are not increased by symmetrization of competing functions. One is then free to search for a minimum among rotationally invariant decreasing functions (which are much easier to analyze since they are essentially functions of a single independent variable). A particular application of this technique has been in the computation of optimal constants for Sobolev inequalities.

Some years ago W. Ni and L. Nirenberg raised the question whether the rearrangement map $\mathcal{R} : f \to f^*$ is strongly continuous in the $W^{1,p}(\mathbf{R}^n)$ topology for all $1 \leq p < \infty$ (this would facilitate application of the «mountain pass lemma», for example). J-M. Coron [CJ] showed such strong continuity (and more) to be true in case $n = 1$, and we, at least, were led to the «obvious» conjecture that continuity holds for all n. We have settled this question [AL] – **rearrangement is not continuous in dimensions larger than one.** As indicated above, we can also identify precisely those f's at which the map $\mathcal{R}$ is continuous and those at which it is not. Our analysis has led us to isolate a property of functions which we call **co-area regularity** which deals with the behavior of functions on their critical sets. For $W^{1,p}$ functions our main result is

THEOREM 1. [AL] *For each $1 \leq p < \infty$ the rearrangement map $\mathcal{R}$ is $W^{1,p}(\mathbf{R}^n)$ continuous at a function f if and only if f is co-area regular.*

Each $W^{1,p}$ function on the line turns out to be necessarily co-area regular so that our theorem is consistent with Coron's result. For higher dimensional domains, however, there are always functions which are not co-area regular. In particular, in $\mathbf{R}^n (n \geq 2)$ there are irregular functions in $C^{n-1,\lambda}$ for each $0 < \lambda < 1$ (i. e. f's which are $n-1$ times continuosly differentiable with $(n-1)^{th}$ derivatives which are Hölder continuous with exponent λ). In fact these irregular functions are dense in $W^{1,p}(\mathbf{R}^n)$. However, each f with Lipschitz $(n-1)^{th}$ derivatives (i.e. $\lambda = 1$) is co-area regular.

In this note we shall briefly review symmetric rearrangement, introduce co-area regularity, sketch the construction of a co-area irregular function, give the reason that co-area irregularity implies lack of continuity of $\mathcal{R}$ in $W^{1,p}$, and finally sketch the reason that co-area regularity implies continuity of $\mathcal{R}$. Our proof of continuity discussed here uses the theory of rectifiable currents in an essential way. The version in [AL] uses more traditional functional analysis instead.

REMARK. One sometimes defines the symmetric decresing rearrangement of vector valued function $f : \mathbf{R}^n \to \mathbf{R}^m$ (as well as functions $\mathbf{R}^n \to \mathbf{R}^+$) by setting $f^* = |f|^*$. Sometimes it is also of interest to replace $W^{1,p}$ norms by gradient energies associated with integrals of other convex integrands $\psi : \mathbf{R}^+ \to \mathbf{R}^+$, i.e. $||\nabla f||_p^p = \int |\nabla f|^p \mathrm{d}\mathcal{L}^n$ is replaced by $\int \psi(|\nabla f|)\mathrm{d}\mathcal{L}^n$. These two generalizations are carried out in [AL] but are omitted here for simplicity. *The conclusions about continuity remain the same.*

It is worth pointing out that although the map $\mathcal{R}$ is not continuous for $W^{1,p}$ norms we show [AL].

THEOREM 2. *For each* $0 < \alpha < 1$, *each* $1 \leq p < \infty$, *and each* $n \geq 1$, *the rearrangement map* $\mathcal{R}$ *is continuous on the fractional Sobolev space* $W^{\alpha,p}(\mathbf{R}^n)$.

For $0 < \alpha < 1$ the norm $||f||_{W^{\alpha,p}}$ is given by

$$||f||_{W^{\alpha,p}}^p = ||f||_p^p + \int\int |f(x) - f(y)|^p\,|x-y|^{-n-p\alpha}\mathrm{d}\mathcal{L}^n x\,\mathrm{d}\mathcal{L}^n y.$$

We have the curious conclusion that co-area regularity plays a role for $W^{\alpha,p}$ only when $\alpha = 1$. Fractional derivatives, of course, are not a local construct.

2. REARRANGEMENTS AND CO-AREA REGULARITY

2.1. Rearrangements

We review the definition and basic properties of the symmetric decreasing rearrangement $f^* = \mathcal{R}f$ of a function $f : \mathbf{R}^n \to \mathbf{R}^+$. It is convenient to use the notation $\chi_{\{A\}} : \mathbf{R}^n \to \{0, 1\}$ symbolically to denote a function which takes value 1 when the test A is passed and takes value 0 otherwise; e.g. $\chi_{\{f>y\}}(x)$ equals 1 in case $f(x) > y$ and equals 0 otherwise. Also we associate to a fixed function f a radius function $R : \mathbf{R}^+ \to \mathbf{R}^+$ defined by requiring

$$\alpha(n) R(y)^n = \int \chi_{\{f>y\}} \mathrm{d}\mathcal{L}^n \tag{2.1}$$

for each y; here $\alpha(n)$ is the volume of the unit ball in $\mathbf{R}^n$. We further denote by $\chi_R : \mathbf{R}^n \to \{0,1\}$ the characteristic function of the open ball centered at the origin and of radius R. Finally, our **rearranged function**

$$f^* : \mathbf{R}^n \to \mathbf{R}^+$$

is defined by setting

$$f^*(x) = \int_{y>0} \chi_{R(y)}(x) \mathrm{d}\mathcal{L}^1 y \tag{2.2}$$

for each x. It is immediate to check that f^* *is symmetric and decreasing,* i.e. $f(x) = f(z)$ if $|z| = |x|$ and $0 \leq f(x) \leq f(z)$ if $|x| \geq |z|$. It is also clear that f^* *is equimeasurable with* f, i.e.

$$\mathcal{L}^n(\{x : f(x) > y\}) = \mathcal{L}^n(\{x : f^*(x) > y\}) \tag{2.3}$$

for each $y > 0$.

Equation (2.3) implies immediately that *rearrangement preserves L^p norms,* i.e.

$$||f||_p = ||f^*||_p. \tag{2.4}$$

Moreover [CG], *rearrangement is a contraction on L^p*, i.e.

$$||f - g||_p \geq ||f^* - g^*||_p, \quad \text{whenever } f, g \in L^p. \tag{2.5}$$

In particular, R *is a continuous map from L^p into L^p.*

The function space $W^{1,p}(\mathbf{R}^n)$ consists of those functions f which belong to $L^p(\mathbf{R}^n)$ and whose distribution gradients ∇f are functions belonging to $L^p(\mathbf{R}^n, \mathbf{R}^n)$. It has long been known [B] [BZ] [H] [K] [L] [PS] [S1] [S2] [T] that $\mathcal{R}$ is $W^{1,p}$ *norm non-increasing,* i.e.

$$||\nabla f||_p \geq ||\nabla f^*||_p. \tag{2.6}$$

This implies that $\mathcal{R}f$ *also belongs to* $W^{1,p}$. (Actually, when $p = 1$ it is not obvious that f^* is in $W^{1,1}$ and not merely in BV; this was proved by Hilden

[H].) However, $\mathcal{R}$ *is not a contraction mapping.* Indeed, $||\nabla f - \nabla g||_p$ can be arbitrarily large compared to $||\nabla f^* - \nabla g^*||_p$. To see why this can happen, suppose that $f, g : \mathbb{R} \to \mathbb{R}^+$ are smooth functions with $f(x) = g(x)$ for $x \leq 0$ and $f(x) > g(x)$ for $x > 0$. Suppose also, for $x \leq 0$, that both ∇f (and hence ∇g) are very large in L^p norm while, for $x > 0$, both ∇f and ∇g are of order 1 in L^p norm. Then $||\nabla f - \nabla g||_p$ is of order one because of the cancellation for $x \leq 0$. On the other hand it is easy to arrange things so that the rearrangement destroys this cancellation so that $||\nabla f^* - \nabla g^*||_p$ will be large.

These facts suggest some of the subtlely of questions about the continuity of $\mathcal{R}$ on $W^{1,p}$. We can phrase our question in the following way.

Given $f, f_1, f_2, \ldots$ *in* $W^{1,p}$ *with* $f_j \to f$ *in* $W^{1,p}$, *is it true that* $A_j = ||\nabla f_j^* - \nabla f^*||_p$ *ultimately converges to 0 as* $j \to \infty$ *even though* A_j *may be large for very many* j*'s?*

2.2. Co-area Regularity

Instead of the integral in (2.1) representing the full crossectional area at height y of the subgraph of our function f, consider the integral

$$\mathcal{G}_f(y) = \int \chi_{\{f>y\}} \chi_{\{\nabla f = 0\}} \mathrm{d}\mathcal{L}^n \tag{2.7}$$

which, for each y, represents that part of the crossection of the subgraph associated with critical points of f. Since our function $\mathcal{G}_f : \mathbb{R}^+ \to \mathbb{R}^+$ is nonincreasing its distribution first derivative G'_f is a (negative) measure. Since a smooth function must be constant on any connected open set on which its gradient vanishes, there are many functions f for which the contributions to the integral in (2.7) come only from flat parts of the graph corresponding to those positive numbers y for which the set $\{x : f(x) = y\}$ has positive measure. Since there can be at most countably many such y's, the measure $\mathcal{G}'_f$ would then be singular with respect to $\mathcal{L}^1$ on $\mathbb{R}^+$. This situation is not the most general one, however, and there are «irregular» smooth functions f for which the measure $\mathcal{G}'_f$ has an absolutely continuous piece as well. Indeed, we have the following theorem.

THEOREM 3. [AL] *For each* $n \geq 2$ *and each* $0 < \lambda < 1$, *there is (by construction) a positive constant* C *and a function* $f : \mathbb{R}^n \to [0,1]$ *withe the following properties.*

(1) The function f *belongs to* $C^{n-1,\lambda}(\mathbb{R}^n)$ *and has support equal to the cube*

$$Q = \{x : |x^i| \leq 1 \text{ for each } i = 1, \ldots, n\}$$

of side length 2.

(2) For each $0 < y \leq 1$,

$$\mathcal{G}_f(y) = C(1 - y).$$

In particular, the measure $\mathcal{G}'_f$ *is absolutely coninuous with respect to* $\mathcal{L}^1$. *Thus* f *is co-area irregular. (See Definition below).*

It can be difficult to picture such a function. Somehow its gradient vanishes on a set of positive $\mathcal{L}^n$ measure containing no open subsets or flat spots, i.e. $\mathcal{L}^n(\{x : f(x) = y\}) = 0$ for every y. Furthermore, the image of the critical set is distributed uniformly over all y values in the range [0, 1].

Theorem 3 also tells us that the following definition is not an empty one.

DEFINITION. *A function* f *in* $W^{1,p}$ *is called* **co-area regular** *if and only if the measure* $\mathcal{G}'_f$ *(see (2.7)) is purely singular with respect to* $\mathcal{L}^1$. *Otherwise* f *is called* **co-area irregular**.

The term co-area in these definitions was suggested by H. Federer's «co-area formula» which gives an integral representation of the absolutely continuous function

$$y \mapsto \int \chi_{\{f>y\}} \chi_{\{\nabla f \neq 0\}} \mathrm{d}\,\mathcal{L}^n.$$

A mild generalization [AL] of the Morse-Sard-Federer theorem shows that each f belonging to $C^{n-1,1}$ is automatically co-area regular. An easy argument then shows

THEOREM 4. [AL] *For each* $n \geq 2$ *and each* $p \geq 1$, *the co-area regular and the co-area irregular functions are each dense in* $W^{1,p}(\mathbf{R}^n)$.

Questions of the behavior of functions on their critical sets have a substantial mathematical heritage both in theory and in examples. We here sketch the construction of a function f as in Theorem 3 when $n = 2$. First set $f(x) = 0$ for $x \notin Q$. For $x \in Q$ we will use 4-adic notation to express the values of our f, i.e. we will write

$$f(x) = \sum_{\ell=1}^{\infty} 4^{-\ell} a_\ell(x) \qquad \text{with } a_\ell(x) \in \{0, 1, 2, 3\}.$$

First divide Q in the obvious way into four squares each of side length 1 and label these squares $S_0^{(1)}, S_1^{(1)}, S_2^{(1)}, S_3^{(1)}$ in clockwise order. Set $a_1(x) = j$ if $x \in S_j^{(1)}$ (don't worry about the boundaries of the $S_j^{(1)}$'s). Next, divide each $S_j^{(1)}$ into four squares each of side length $\frac{1}{2}$ and label these $S_{jk}^{(2)}$ (with $k = 0, 1, 2, 3,$) in the same clockwise order. Set $a_2(x) = k$ if $x \in S_{jk}^{(2)}$. The construction continues in the obvious way ulimately to define an f. For each $0 < a < b < 1$ we have $\mathcal{L}^2(f^{-1}(a,b)) = 4(b-a)$. At present our f is not even continuous much less smooth. We fix this up by modifying this construction. We replace each a_ℓ by a carefully constructed smooth function b_ℓ in our sum above. The support of each b_ℓ is contained within the $4^{\ell-1}$ squares on which $b_{\ell-1}$ assumes constant values, and. b_ℓ assumes constant values on 4^ℓ squares nested within the $b_{\ell-1}$ constant value squares. The subgraph then resembles a union of step pyramids (like Zhoser not Cheops) with those at the ℓ-th level having bases on the tops of those at the $\ell-1$-th level. With some effort one can construct the b_ℓ's so that $f \in C^{1,\lambda}$ and $\{x : \nabla f = 0\}$ has positive measure. As expected the measure of the set $\{x : \nabla f = 0\}$ goes to zero as λ approaches 1.

3. REARRANGEMENT IS DISCONTINUOUS AT CO-AREA IRREGULAR FUNCTIONS

THEOREM 5. [AL] *Suppose $n \geq 2$ and f is a co-area irregular function belonging to $W^{1,p}(\mathbb{R}^n)$. Then there is a sequence $f_1, f_2, f_3, \ldots$ of functions in $W^{1,p}(\mathbb{R}^n)$ such that $f_j \to f$ in $W^{1,p}(\mathbb{R}^n)$ as $j \to \infty$ but $f_j^* \not\to f^*$. Moreover, for each $\epsilon > 0$, the f_j's can be chosen with the following properties.*

(1) The sequence of differences $f_j - f$ converges to zero in $L^\infty(\mathbb{R}^n)$.

(2) There is a positive number Y such that

$$f_j(x) = f(x) \qquad \text{whenever} \quad f(x) < Y \text{ or } f(x) > Y + \epsilon$$

and

$$Y \leq f_j(x) \leq Y + \epsilon \qquad \text{whenever} \quad Y \leq f(x) \leq Y + \epsilon.$$

(3) For $\mathcal{L}^n$ almost every x,

$$|\nabla f_j(x)| \leq \frac{3}{2}|\nabla f(x)| + \epsilon$$

(4) The measure of the set

$$\{x : \nabla f(x) \neq 0 \text{ and } \nabla f_j(x) \neq \nabla f(x)\}$$

converges to zero as $j \to \infty$.

If we do not require properties (2), (3), (4) then the difference $f_j - f$ can be chosen to belong to C^∞. If we drop all four properties then each f_j can be chosen to belong to C^∞. The basic idea behind the proof of Theorem 5 (omitting refinements

(1), (2), (3), (4)) is the following. Let W be the characteristic function of the critical set of f, i.e. the set for which $\nabla f = 0$, and set

$$f_j(x) = f(x) + \frac{1}{2j} W(x) \sin(jf(x)) \tag{3.1}$$

for each x. Then clearly $f_j \to f$ in L^p as $j \to \infty$. For the gradients we compute formally

$$\begin{aligned} \nabla f_j(x) - \nabla f(x) = \frac{1}{2} W(x) \nabla f(x) \cos(jf(x)) \\ + \frac{1}{2j} \nabla W(x) \sin(jf(x)). \end{aligned} \tag{3.2}$$

The first term on the right side is zero since W vanishes when ∇f does not vanish. The second term on the right side in (3.2) is a bit problematic since ∇W is not p-th power summable. This defect, however, can be remedied (with some effort) by mollifying W in a j-dependent way so that $\|\nabla W\|_\infty < j^{1/2}$. This establishes the L^p convergence of ∇f_j to ∇f.

Now define sets

$$K_j(y) = \{x : f_j(x) > y\} \quad (j = 1, 2, 3, \ldots) \quad \text{and} \quad K(y) = \{x : f(x) > y\}$$

for each y. Since the function $t \mapsto t + \frac{1}{2j}\sin(tj)$ is increasing (check the derivative) we infer that $K_j(y) = K(y)$ whenever m is an integer and $y = 2m\pi/j$. For these special y values we infer that the radius functions are equal, i.e. $R_j(y) = R(y)$ (recall (2.1)). On the other hand, if $0 < \sigma < 1$ and $y = (2m + \sigma)(\pi/j)$, then $R_j(y) \geq R(y)$ and, in general, $R_j(y) \geq R(y)$.

Think of the graphs of f_j^* and f^* parametrized by the height y instead of the radius $|x|$. When $y = 2m\pi/j$ the graphs intersect. When $y = (2m+\sigma)(\pi/j)$ and $0 < \sigma < 1$, the graph of f_j^* lies to the right of the graph of f^*. For our purposes it sufficies to show that the numbers $B_j \equiv \|\nabla f_j^* - \nabla f^*\|_1$ are bounded away from zero. We then try to estimate the B_j's in terms of the distribution $\mathcal{G}_f$ from (2.7). Using the Schwarz inequality several times and a simple Sobolev inequality we are able to estimate

$$B_j \geq (\text{constant}) \int |h|^{1/2} \mathrm{d}\,\mathcal{L}^1; \tag{3.3}$$

here $\mathcal{L}^1 \wedge h$ denotes the absolutely continuous part of our $\mathcal{G}_f'$.

It is reassuring that the bound (3.3) above involves $|h|^{1/2}$ instead of $|h|$. This is so because «the square root of a *singular* measure is zero»; by this we mean that if the singular part of $\mathcal{G}_f'$ (which cannot contribute to the lack of convergence, as we assert in the next section) is approximated by absolutely continuous measures $\mathcal{L}^1 \wedge \tilde{h}^{(k)}$ $(k = 1, 2, 3, \ldots)$, then $\int |\tilde{h}^{(k)}|^{1/2} \mathrm{d}\,\mathcal{L}^1$ converges to zero as $k \to \infty$.

4. REARRANGEMENT IS CONTINUOUS AT CO-AREA REGULAR FUNCTIONS

The proof [AL] that the co-area regularity of f implies $W^{1,p}$ continuity of $\mathcal{R}$ at f is quite technical. We will attempt to outline some of the main ideas. In our proof in [AL] sections 4.2 and 4.3 below are replaced by more traditional methods in functional analysis.

4.1. Reduction to $W^{1,1}$

Our first step is to establish the fact that continuity of $\mathcal{R}$ in $W^{1,p}$ is implied by continuity of $\mathcal{R}$ in $W^{1,1}$. This may seem surprising since ordinarily nothing can be inferred about $||\nabla f_j^* - \nabla f^*||_p$ from information about $||\nabla f_j^* - \nabla f^*||_1$. In the present case, however, our rearrangement operator $\mathcal{R}$ acts independently on slabs $\{x : Y_1 < f(x) < Y_2\}$. We can then surgically remove small, well chosen slabs from the f_j and f on which $|\nabla f_j^*|$ or $|\nabla f^*|$ is large. On these slabs we can control $||\nabla f_j^* - \nabla f^*||_p$ in terms of $||\nabla f_j^*||_p$ and $||\nabla f^*||_p$ and these quantities can, in turn, be controlled by $||\nabla f_j||_p$ and $||\nabla f||_p$ with use of the basic inequality (2.6). After these small slabs are removed, the f_j and f effectively have bounded gradients and then $W^{1,1}$ convergence implies $W^{1,p}$ convergence.

4.2. The co-area formula and co-area regularity

The basic tool in our second step is H. Federer's co-area formula as extended by J. Brothers and W. Ziemer [BZ]. Suppose $f \in W^{1,1}(\mathbf{R}^n)$ and g is a nonnegative Borel function. Then the slice integral

$$A(y) \equiv \int_{f^{-1}\{y\}} g \, \mathrm{d}\mathcal{H}^{n-1} \tag{4.1}$$

exists for $\mathcal{L}^1$ almost every positive number y and we have the *co-area formula*

$$\int_{y>0} A \, \mathrm{d}\mathcal{L}^1 = \int g|\nabla f| \, \mathrm{d}\mathcal{L}^n; \tag{4.2}$$

here $\mathcal{H}^{n-1}$ denotes Hausdorff's $(n-1)$-dimensional measure over $\mathbf{R}^n$. In one application of (4.2) we replace $f(x)$ by $F_t(x) = \max\{f(x), t\}$ (with $t > 0$), then replace $g(x)$ by $(|\nabla f(x)| + \delta)^{-1}$, then let $\delta \to 0+$, and finally use Lebesgue's monotone convergence theorem applied to each side of (4.2) to infer

$$\int_{y>0} \omega_f(y) \, \mathrm{d}\mathcal{L}^1 y = \int \chi_{\{f>t\}} \chi_{\{\nabla f \neq 0\}} \mathrm{d}\mathcal{L}^n \equiv \gamma_f^{(t)} \tag{4.3}$$

where we have written

$$\omega_f(y) = \int_{f^{-1}\{y\}} |\nabla f|^{-1} \mathrm{d}\mathcal{H}^{n-1} \tag{4.4}$$

for each y. In other words, the basic distribution integral on the right side of (2.1) (call it $\sigma_f(y)$) breaks up naturally into two pieces

$$\sigma_f(y) = \gamma_f(y) + \mathcal{G}_f(y) \tag{4.5}$$

and (4.3) states that γ_f is absolutesly continuous with derivative $-\omega_f$. The KEY POINT is: *the only absolutely continuous part of the measure* $-\sigma'_f$ *is* ω_f *if and only if* f *is co-area regular.*

4.3. Currents and the lower semicontinuity of slice integrals

Suppose that we have a sequence f_j converging to f in $W^{1,1}$ and that f is co-area regular. Henceforth we will omit the subscript f (e.g. σ_f will be denoted σ) when referring to f, and will use a subscript j when referring to f_j (e.g. σ_{f_j} will be denoted σ_j). We assert that

$$\liminf_{j\to\infty} \omega_j(y) \geq \omega(y) \qquad \mathcal{L}^1 \text{ almost every } y. \tag{4.6}$$

To show this it sufficies to prove that

$$\lim_{j\to\infty} \int_{f_j^{-1}\{y\}} g \,\mathrm{d}\mathcal{H}^{n-1} = \int_{f^{-1}\{y\}} g \,\mathrm{d}\mathcal{H}^{n-1} \qquad \text{for } \mathcal{L}^1 \text{ almost every } y \tag{4.7}$$

whenever $g \in L^\infty$. An approximation argument shows it is sufficient to prove (4.7) for $g \in C_0^\infty$. It is here that we need to utilize the inherent current structure of the graph and subgraph of f and the f_j's and the inherent convergence as currents. To do this we form the $n+1$ dimensional current

$$Q = E^{n+1} \llcorner \{(x,y) : x \in \mathbf{R}^n, y < f(x)\}$$

whose boundary $T = \partial Q$ is the current associated with the graphs of f. The current T can then be sliced by the coordinate function y to obtain an $n-1$ dimensional slice current $T(y)$ corresponding to the level set f^{-1} for $\mathcal{L}^1$ almost every y. Likewise, we define $Q_j, T_j, T_j(y)$ for the various j's and further set $S_j = Q - Q_j$ with associated slice currents $S_j(y)$. Since «slicing commutes with boundaries» in the current setting we infer $\partial S_j(y) = T(y) - T_j(y)$ for almost every y.

Since the mass M of a current corresponds to its volume, we readily check that

(4.8) $$M(S_j) = M(Q - Q_j) = \|f - f_j\|_1 \to 0 \qquad \text{as } j \to \infty.$$

Since $M(S_j) = \int M(S_j(y)) \mathrm{d}\mathcal{L}^1 y$ for each j, there will be a subsequence (still denoted by j's) such that

(4.9) $$\lim_{j\to\infty} M(S_j(y)) = 0 \qquad \text{for almost every } y.$$

Since $\partial S_j(y) = T(y) - T_j(y)$ we conclude the convergence of the $T_j(y)$'s to $T(y)$ for almost every y. The lower semicontinuity of mass under such convergence then implies

(4.10) $$\liminf_{j\to\infty} M(T_j(y)) \geq M(T(y)) \quad \text{for } \mathcal{L}^1 \text{ almost every } y.$$

Using, for example, J. Michael's [M] Lipschitz approximation theorem we readily infer

(4.11) $$M(T_{(j)}(y)) = \mathcal{H}^{n-1}(f_{(j)}^{-1}(y)) \qquad \text{for } \mathcal{L}^1 \text{ almost every } y;$$

here (j) denotes either j or no j. We use the co-area formula again to infer

(4.12) $$\int M(T_{(j)}(y)) \mathrm{d}\mathcal{L}^1 y = \int |\nabla f_{(j)}| \mathrm{d}\mathcal{L}^n.$$

However, $\int |\nabla f_j| \mathrm{d}\mathcal{L}^n \to \int |\nabla f| \mathrm{d}\mathcal{L}^n$ by the assumed L^1 convergence of ∇f_j to ∇f.

The following is a general lemma. Suppose μ is a measure and $h, h_1, h_2, h_3, \ldots$ are nonnegative, summable functions such that $\liminf_j h_j(x) \geq h(x)$ for μ almost every x. In case $\int h_j \mathrm{d}\mu \to \int h \mathrm{d}\mu$ as $j \to \infty$ then there is a subsequence $j(k)$ of the j's such that $h_{j(k)}(x) \to h(x)$ as $k \to \infty$ for μ almost every x.

We apply this lemma to the case at hand to infer that, for a further subsequence,

(4.13) $$\liminf_{j\to\infty} M(T_j(y)) = M(T(y)) \qquad \text{for } \mathcal{L}^1 \text{ almost every } y.$$

Equation (4.13), with a little more work, then leads to (4.7).

As an application of (4.7) we return to (4.4) and prove that

(4.14) $$\liminf_{j\to\infty} \omega_j(y) \geq \omega(y) \qquad \text{for } \mathcal{L}^1 \text{ almost every } y.$$

This result is crucial for us. To prove it, we use (4.7) with $g_{(j)}(y) = (|\nabla f_{(j)}| + \delta)^{-1}$ (as in the proof of (4.4)) and then let $\delta \to 0$.

4.4. Graph arc length as an invariant measure

The last main step in our proof is to combine (4.14), the co-area regularity of f, and the $W^{1,1}$ convergence of the f_j's to f to show that the ∇f_j^*'s convergence to ∇f^* in L^1. Since $f_{(j)}^*(x)$ is really only a function of $r = |x|$, our considerations are essentially one-dimensional. (It is true that the real measure is $r^{n-1}\,dr$ and not dr, but this is merely a nuisance which one can handle). Let us suppose then $n = 1$ and we will denote d/dr by a prime.

Think of the graph of f^* (or f_j^*) which is a curve in $\mathbb{R}^2$. The geometrically invariant notion is not $f^{*\prime}$ (which is the quantitativity in which we are really interested) but rather the arc length derivative $(1 + (f^{*\prime})^2)^{1/2}$. The arc length can be computed in two different ways. The first way is to use the height y as parameter. Then the arc length of the graph of $f_{(j)}^*$ equals

$$\int (1 + (\rho_{(j)}(y))^2)^{\frac{1}{2}} \mathrm{d}\mathcal{L}^1 y + \int \mathrm{d}\nu_{(j)};$$

here $\nu_{(j)}$ is the singular part of the measure $-(\mathrm{d}\sigma_{(j)}/\mathrm{d}y)$ while $\mathcal{L}^1 \wedge \rho_{(j)}$ is the absolutely continuous part of $-(\mathrm{d}\sigma_{(j)}/\mathrm{d}y)$. *The crucial point is the following: The co-area regularity of f implies that $\rho(y) = \omega(y)$.* For f_j^*, all we can say is that $\rho_j(y) \geq \omega_j(y)$; but this is of no concern since, from (4.14), we have

$$\liminf_{j\to\infty} \rho_j(y) \geq \rho(y) \qquad \text{for } \mathcal{L}^1 \text{ almost every } y. \tag{4.15}$$

Concerning the singular components ν_j and ν one knows nothing. However, by the L^1 convergence of f_j^* to f^* (see (2.5)) we can infer that the arcs convergence pointwise, i.e. for any $0 < a < b$

$$\int_a^b \rho_j \mathrm{d}\mathcal{L}^1 + \int_a^b \mathrm{d}\nu_j \to \int_a^b \rho \mathrm{d}\mathcal{L}^1 + \int_a^b \mathrm{d}\nu. \tag{4.16}$$

It is then a simple exercise to show that (4.15), (4.16) alone imply *arc length convergence*, i.e.

$$\int (1 + \rho_j^2)^{1/2} \mathrm{d}\mathcal{L}^1 + \int \mathrm{d}\nu_j \to \int (1 + \rho^2)^{1/2} \mathrm{d}\mathcal{L}^1 + \int \mathrm{d}\nu. \tag{4.17}$$

Now think about this arc length convergence (4.17) in terms of the radius parameterization, i.e.

$$\int (1 + (f_j^{*\prime}(r))^2)^{1/2} \mathrm{d}\mathcal{L}^1 r \to \int (1 + (f^{*\prime}(r))^2)^{1/2} \mathrm{d}\mathcal{L}^1 r. \tag{4.18}$$

There is no singular part of the measure (since $f_j^{*\prime}$ is a function). Intuitively, it is clear (by drawing a few graphical examples) that arc length convergence implies L^1 convergence of $f_j^{*\prime}$ to $f^{*\prime}$ because the function $t \mapsto (1 + t^2)^{1/2}$ is strictly convex. This is indeed correct as the following general theorem [AL] states.

THEOREM 6. *Suppose* $\psi : \mathbf{R}^n \to \mathbf{R}^+$ *is a convex function. Suppose also that* $f, f_1, f_2, f_3, \ldots$ *are functions in* $L^1_{\text{loc}}(\mathbf{R}^n, \mathbf{R})$ *having distributional gradients which are functions in* $L^1_{\text{loc}}(\mathbf{R}^n, \mathbf{R})$. *Suppose that* $\psi(\nabla f), \psi(\nabla f_1), \psi(\nabla f_2), \psi(\nabla f_3), \ldots$ *also are functions in* $L^1(\mathbf{R}^n)$ *and that* $f_j - f \to 0$ *in* $L^1(\mathbf{R}^n)$ *as* $j \to \infty$. *Then (as has been known for some time [SJ])*

$$\text{(1)} \qquad \liminf_{j\to\infty} \int \psi(\nabla f_j)\,\mathrm{d}\mathcal{L}^n \geq \int \psi(\nabla f)\,\mathrm{d}\mathcal{L}^n.$$

(2) Suppose further that equality holds in (1) and that ψ *is strictly convex (i.e.* $\psi(x) + \psi(y) > 2\psi\left(\frac{x+y}{2}\right)$ *whenever* $x \neq y$). *Uniform convexity is not assumed. Then* $\psi(\nabla f_j) \to \psi(\nabla f)$ *in* $L^1(\mathbf{R}^n)$ *as* $j \to \infty$. *Furthermore, there is a subsequence* $j(1), j(2), j(3), \ldots$ *of* $1, 2, 3, \ldots$ *such* $\nabla f_{j(k)}(x) \to \nabla f(x)$ *for* $\mathcal{L}^n$ *almost every* x *as* $k \to \infty$.

(3) Finally, suppose $\psi(\xi) \to \infty$ *as* $|\xi| \to \infty$ *(e.g. our function* $\xi \mapsto (1 + |\xi|^2)^{1/2}$). *Then, for every measurable subset* Ω *of* $\mathbf{R}^n$ *of finite measure,* $\nabla f_j \to \nabla f$ *in* $L^1(\Omega, \mathbf{R}^n)$.

REFERENCES

[AL] F. ALMGREN and E. LIEB: Symmetric decresing rearrangement is sometimes continuous, J. Amer. Math. Soc. **2**, 683-773 (1989).

[B] C. BANDLE: Isoperimetric inequalities and applications. Pitman (Boston, London, Melboune), 1980.

[BZ] J. BROTHERS and W. ZIEMER: Minimal rearrangements of Sobolev functions. *Journ. Reine Angew. Math.* **384**, 153-179 (1988).

[CG] G. CHITI: Rearrangements of functions and convergence in Orlicz spaces. *Appl. Anal.* **9**, 23-27 (1979).

[CJ] J-M. CORON: The continuity of the rearrangement in $W^{1,p}(\mathbb{R})$. *Ann. Scuol. Norm. Sup. Pisa, Ser 4*, **11**, 57-85 (1984).

[H] K. HILDEN: Symmetrization of functions in Sobolev spaces and the isoperimetric inequality. *Manuscr. Math.* **18**, 215-235 (1976).

[K] B. KAWOHL: Rearrangements and convexity of level sets in partial differential equations. *Lect. Notes in Math.* **1150**, Springer (Berlin, Heidelberg, New York), 1985.

[L] E. LIEB: Existence and uniqueness of the minimizing solution of Choquard's nonlinear equation. *Stud. Appl. Math.* **57**, 93-105 (1977). See appendix.

[M] J. MICHAEL: Lipschitz approximations to summable functions. *Acta Math.* **111**, 73-94 (1964).

[PS] G. POLYA and G. SZEGÖ: Isoperimetric inequalities in mathematical physics. *Ann. Math. Stud.* **27**, Princeton University Press (Princeton) (1951).

[SJ] J. SERRIN: On the definition and properties of certain variational integrals. *Trans. Amer. Math. Soc.* **101**, 139-167, (1961).

[S1] E. SPERNER: Zur symmetrisierung von Funktionen auf Sphären. *Math. Z.* **134**, 317-327 (1973).

[S2] E. SPERNER: Symmetrisierung für Funktionen mehrerer reeller Variablen. *Manuscr. Math.* **11**, 159-170 (1974).

[T] G. TALENTI: Best constant in Sobolev inequality. *Ann. Pura Appl.* **110**, 353-372 (1976).

COUNTING SINGULARITIES IN LIQUID CRYSTALS

Frederick J. Almgren Jr. - Elliott H. Lieb

Abstract. *Energy minimizing harmonic maps from the ball to the sphere arise in the study of liquid crystal geometries and in the classical nonlinear sigma model. We linearly dominate the number of points of discontinuity of such a map by the energy of its boundary value function. Our bound is optimal (modulo the best constant) and is the first bound of its kind. We also show that the locations and numbers of singular points of minimizing maps is often counterintuitive; in particular, boundary symmetries need not be respected.*

1. INTRODUCTION

This note is an introduction to and summary of discoveries we have made about the singular behaviour of

- A mathematical model of some liquid crystal geometries
- Dirichlet energy minimizing harmonic maps from regions in $\mathbf{R}^3$ to S^2
- Energy minimizing configurations of a classical nonlinear sigma model

$(R^3 \to S^2)$.

These phenomena are different facets of a common mathematical analysis set forth in detail in our paper [AL]. There we study vector fields φ of unit length defined in a reasonable region Ω in $\mathbf{R}^3$. In coordinates we can thus write for each $x = (x^1, x^2, x^3)$ in Ω,

$$(1) \qquad \varphi(x) = (\varphi^1(x), \varphi^2(x), \varphi^3(x)) \qquad \text{with} \qquad \sum_{i=1}^{3} \varphi^i(x)^2 = 1.$$

Since our target S^2 is 2-dimensional we could, in principle, describe φ using two functions instead of our three constrained functions. It is easier, however, to work with three functions and a constraint.

The φ's important for us have distribution first derivatives which are square summable. (Caution: the space of such φ's satisfying (1) is not the completion of any space of smooth mappings $\Omega \to S^2$.) The gradients of such φ's are defined for almost every x with norms represented by the formula

$$|\nabla\varphi(x)|^2 = \sum_{i=1}^{3}\sum_{\alpha=1}^{3}\left(\frac{\partial\varphi^i(x)}{\partial x^\alpha}\right)^2$$

which gives the value of **Dirichlet's integrand** at x. The integral of this integrand is **Dirichlet's energy integral** of φ,

$$\mathcal{E}(\varphi) = \int_\Omega |\nabla\varphi|^2 \mathrm{d}V,$$

with $\mathrm{d}V = \mathrm{d}x^1\mathrm{d}x^2\mathrm{d}x^3$. Critical points of this energy integral $\mathcal{E}$ are by definition **harmonic functions** and satisfy the associated **Euler-Lagrange partial differential equations**

$$-\Delta\varphi^i(x) = \varphi^i(x)|\nabla\varphi(x)|^2 \qquad (i = 1,2,3).$$

These equations state that a critical φ has vanishing Laplacian in directions in which it is unconstrained. Such an energy functional and associated partial differential equations appear in the physics literature under the rubric of **the nonlinear sigma model**.

Somewhat more generally, reasonable maps $\varphi : \mathcal{M} \to \mathcal{N}$ between Riemannian manifolds $\mathcal{M}$ and $\mathcal{N}$ (often submanifolds of Euclidean vector spaces) have a Dirichlet's energy integral

$$\mathcal{E}_{\mathcal{M}\mathcal{N}}(\varphi) = \int_\mathcal{M} |D\varphi|^2 \mathrm{d}V_\mathcal{M}$$

of which ours is a special case. Alternatively, one can write

$$\mathcal{E}_{\mathcal{M}\mathcal{N}}(\varphi) = \int_\mathcal{M} g_{ij}(\varphi(x))G^{\alpha\beta}(x)\left(\frac{\partial\varphi^i}{\partial x^\alpha}(x)\right)\left(\frac{\partial\varphi^j}{\partial x^\beta}(x)\right)\mathrm{d}V_\mathcal{M}x$$

where g is the metric on N, G is the metric on M and $\mathrm{d}V_\mathcal{M}x = (\det G(x))^{1/2}$ $\mathrm{d}x$. Extremal mappings for such energies are also called **harmonic mappings**. Such mappings often are not continuous and there in an extensive mathematical theory about them.

The φ's mapping Ω to S^2 which are important for us also have well defined **boundary functions** $\psi : \partial\Omega \to S^2$ having **boundary energy**

$$\partial\mathcal{E}(\psi) = \int_{\partial\Omega} |\nabla_T \psi|^2 \mathrm{d}\, A$$

which is finite; here $\nabla_T \psi$ is the tangential gradient of ψ and $\mathrm{d}\, A$ is surface area measure. Associated with such a ψ is the number

$$E(\psi) = \inf\ \{\mathcal{E}(\varphi) : \varphi \text{ has boundary value function } \psi\}.$$

We call φ an **energy minimizing map** for boundary value function ψ if and only if

$$\mathcal{E}(\varphi) = E(\psi).$$

If Ω is any reasonable bounded domain and ψ is any boundary value function of finite energy then there will always be at least one minimizer φ having ψ as boundary values (a compactness argument). Sometimes, however, there can be more than one minimizer. This is one of the fascinations of this simple nonlinear problem; if the target S^2 were replaced by $\mathbb{R}^3$ (i.e. our constaint were removed) then the Euler-Lagrange partial differential equations are (the unconstrained) linear partial differential equations of Laplace, $\Delta\varphi^i = 0\ (i = 1, 2, 3)$, for which uniqueness is well known.

If our domain Ω is all of $\mathbb{R}^3$ there is no boundary value function ψ, of course. We then say that $\varphi : \mathbf{R}^3 \to S^2$ is a minimizer provided φ cannot be modified on a compact set K to decrease energy in a larger bounded open set containing K.

Liquid crystals

The connection of our energy minimizing φ's with liquid crystals requires explanation. We imagine that Ω is a container containing a liquid crystal. At points x in Ω the liquid determines a **directrix** $n(x)$ lying in real projective space $\mathbf{RP}^2$. Since $\mathbb{RP}^2$ is obtained from S^2 by identifying antipodal points, this means intuitively that $n(x)$ is a unit vector like our $\varphi(x)$ except that its head is indistinguishable from its tail. For the liquid crystals with which we are concerned, the energy of n is defined analogously to our $\mathcal{E}$, e.g. zero energy corresponds to parallel alignment. Like our minimizing φ's (as we shall see), any minimizing n will be continuous except at isolated points. This means, in particular, that any minimizing n can locally be lifted to become a minimizing φ having the same energy; this lifting is global in case Ω is simply connected. (see [BCL], p. 686 for details). Thus, for simply connected Ω's, our original problem is equivalent to the liquid crystal problem. In any case, whether or not Ω is simply connected, our estimates on the number of singular points hold for these liquid crystal minimizers. Line singularities do not occur in our model because they would have infinite Dirichlet energy. They do occur in nature, but to model them one, effectively, has to fatten the line and treat it separately (much as in the liquid helium problem). A further complication for liquid crystals is that there are other, more appropriate, integrands which are quadratic in $\nabla\varphi$ and respect rotational symmetry. The general nematic liquid crystal integrand, for example, is of the form

$$K_1(\operatorname{div}\varphi)^2 + K_2(\varphi\cdot\operatorname{curl}\varphi)^2 + K_3(\varphi\wedge\operatorname{curl}\varphi)^2.$$

Our Dirichlet's energy integrand corresponds (except for a fixed boundary term) to setting $K_1 = K_2 = K_3 = 1$ (see [BCL], p. 653). Our methods give information about such liquid crystal geometries (by a compactness argument) only when K_1, K_2 and K_3 are nearly equal.

2. BASIC FACTS ABOUT MINIMIZERS

(A) Existence and regularity of minimizers

As we mentioned above, whenever we have a reasonable domain Ω and boundary function ψ of finite energy, there will always exist a minimizer φ having boundary values ψ. Such a result is included among the general analysis of Dirichlet's integral minimizing mappings between manifolds by R. Schoen and K. Uhlenbeck in their basic papers [SU1] [SU2]. They further showed that a minimizing φ in our context is a real analytic mapping except at isolated points of discontinuity (which are our singularities). Finally, they concluded that a minimizing φ assumes its boundary values smoothly when both $\partial\Omega$ and ψ are comparably smooth.

(B) Monotonicity of energy and tangential approximations

One of the basic technical properties of energy minimizing mappings is usually called **monotonicity**. Whenever φ is a minimizer in $\Omega, y \in \Omega$, and $0 < r < s < R$ so that the ball $B_R(y)$ also lies within Ω, then

$$\frac{1}{r}\int_{B_r(y)} |\nabla\varphi|^2 \mathrm{d}V \le \frac{1}{s}\int_{B_s(y)} |\nabla\varphi|^2 \mathrm{d}V.$$

For a proof, see [SU1]. (The absence of a corresponding monotonicity estimate is the main reason our analysis of liquid crystals is restricted to the $K_1 = K_2 = K_3$ case). The monotonicity estimate leads fairly directly to the existence of certain tangential approximations to φ at each interior y. A major and deep development occured in a paper of L. Simon [S] which for our problem guarantees the existence of a **unique tangential approximating mapping.** At regular points this approximating mapping is constant. For a singular point y of φ in Ω, Simon's result gives a **unique harmonic mapping** $f : S^2 \to S^2$ such that

$$\varphi(y + t\omega) \to f(\omega) \qquad \text{as} \qquad t \to 0+$$

uniformly for all ω's in S^2 (see [AL]), i.e.

$$\varphi(x) \approx f\left(\frac{x-y}{|x-y|}\right)$$

for x's near y. The correspondence here is in several strong senses (see [AL]). In general, if $f : S^2 \to S^2$ and $F : \mathbf{R}^3 \to S^2$ is defined by setting

$$F(x) = f\left(\frac{x}{|x|}\right)$$

for each $x \neq 0$ then f is harmonic if and only if F is.

EXAMPLE. $f\left(\frac{x}{|x|}\right) = \frac{x}{|x|}$, i.e. f is the identity; see Figure 1.

(C) Harmonic mappings between spheres and mapping degrees

Any continuous mapping $S^2 \to S^2$ has a well defined topological degree measuring the number of times the first sphere covers the second, taking into account the orientations. Since the boundary functions ψ under consideration map S^2 to S^2 and have finite energy, they also have a well defined degree given by the Jacobian integral

$$\deg(\psi) = \frac{1}{4\pi}\int_{\partial\Omega} J(\psi)\,\mathrm{d}A;$$

here $J(\psi)$ is the Jacobian (determinant) function of ψ whose sign is positive or negative at a point depending on whether $D\psi$ preserves or reverses orientations at that point. For continuous ψ's of finite energy these two notions of degree coincide.

All possible harmonic mappings from S^2 to S^2 have been classified for some time. In complex coordinates (resulting from stereographic projection of the S^2's onto $\mathbb{C}$) they are all of the form

$$f(z) = \frac{P(z)}{Q(z)} \qquad \text{or} \qquad f(z) = \frac{P(\bar{z})}{Q(\bar{z})}$$

corresponding to various complex polynomial functions P and Q which are relatively prime. The degree of these f's can be checked to be

$$\deg(f) = \begin{cases} \max(\deg(P), \deg(Q)) & \text{first case;} \\ -\max(\deg(P), \deg(Q)) & \text{second case.} \end{cases}$$

For these harmonic maps $f : S^2 \to S^2$ we also set $F(x) = f\left(\frac{x}{|x|}\right)$ as above and compute for each $0 < R < \infty$ that

$$\int_{|x|<R} |\nabla F|^2 \mathrm{d}V = 8\pi R |\deg(f)|,$$

i.e. the energy does not depend on P and Q except via the degree.

(D) Tangential approximations to minimizers

Suppose $y \in \Omega$ is a singular point of a minimizer φ and the tangential approximation is of the form $F(x) = f\left(\frac{x}{|x|}\right)$ corresponding to one of the harmonic f's given in (C) above. By the **degree of the singular point** y we mean the mapping degree of the associated f. **Which of the possible f's actually occur?** This question was answered by H. Brezis, J-M. Coron, and E. Lieb in their paper [BCL]. The *only* f's that occur are rotations $\mathcal{R}$ and reflections of the f in the above example, i.e.

$$(2) \qquad f(\omega) = \pm\mathcal{R}(\omega), \quad (\omega \in S^2) \qquad \text{with} \quad \deg(f) = \pm 1;$$

see Figure 1. This class does not even include all harmonic maps of degree ± 1. The proof proceeds by a construction of comparison functions. If $|\deg(f)| > 1$ then the energy of F can be decreased by splitting the singularity at the origin into two nearby singularities of lower degree. If $|\deg(f)| = 1$ and $f \neq \pm\mathcal{R}$ then the energy of F can be decreased by moving the singular point slightly.

The paper [BCL] also answered a question that in some sense is complementary to the minimization question we have been studying here. Suppose $y_1, \dots y_n$ are *fixed* points in Ω and $\mathrm{d}_1, \dots, \mathrm{d}_n$ are *fixed* degrees associated to these points (not necessarily ± 1).

What is the infimum of energies $\mathcal{E}(\varphi)$ among all φ's which are continuous except at y_i's and map small spheres around each y_i with degree d_i?

The boundary function ψ is not fixed. This infimum is *not* achieved in general. The answer is shown in the Figure 2. Think of each singularity as a source or sink of flux and draw lines to carry the flux between singularities, or between a singularity and the boundary. Then

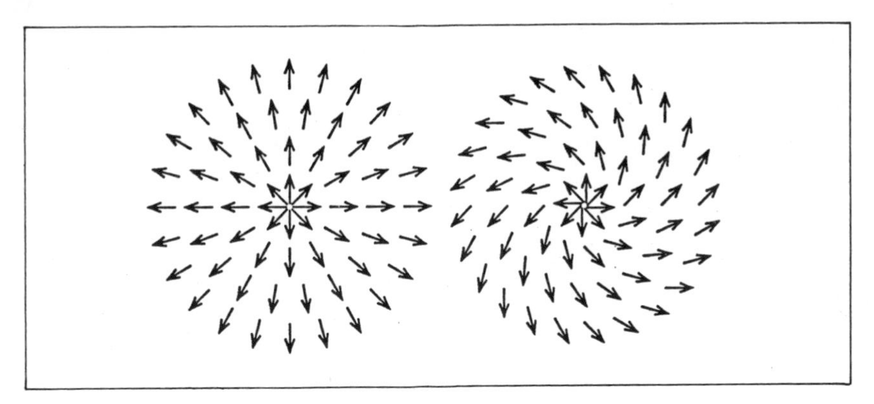

Fig. 1. Here are shown representations of unit vector fields

$$F(x) = \left(\frac{x}{|x|}\right) \quad \text{and} \quad G(x) = \mathcal{R}\left(\frac{x}{|x|}\right)$$

in which $\mathcal{R}$ is a counterclockwise rotation through $45°$. Such arrays minimize Dirichlet's integral energy and are also observed as stable liquid crystal geometries [K].

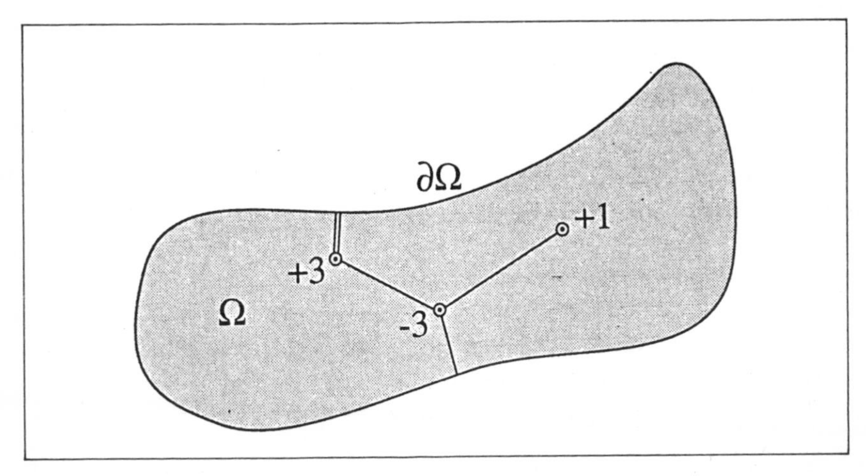

Fig. 2. A region Ω is pictured here containing three prescribed singular points whose degrees $(+3, -3, +1)$ are also prescribed. The least energy of unit vector fields having this singular behavior is the least total mass of oriented line segments connecting these singular points (as currents) either to each other or to the boundary. Such a least length array is illustrated.

$$\inf \mathcal{E}(\varphi) = 8\pi \min \left\{ \sum \text{lengths of lines} \right\}$$

where the minimum is over all ways of constructing the lines. A different proof of this result was later given by F. Almgren, W. Browder, and E. Lieb [ABL] using H. Federer's co-area formula in the context of currents. This is like quark confinement: a plus and minus quark have an energy proportional to their separation.

From this result with *specified* singularities one is tempted to surmise that, in our original minimization problem, *potential* singularities would tend to annihilate each other (if of opposite degrees) or move to $\partial\Omega$. The number of singularities that will occur will be only that required by topology, i.e.

$$\sum_{\text{singularities}} \deg(\text{singularity}) = \deg(\psi) = \frac{1}{4\pi} \int_{\partial\Omega} J(\psi)\, \mathrm{d}A.$$

This surmise is very wrong, as we shall see later in Example 3, and misled us for a long time. Arbitrarly many singularities (of mixed signs) can occur, even if the Jacobian $J(\psi)$ vanishes identically.

(E) Boundary regularity and hot spots

Our main estimates require an extension of the boundary regularity results indicated above in (A). These theorems take several pages merely to state precisely, but the essence of the matter is the following. Assume that $\partial\Omega$ is smooth and take a small patch $P \subset \partial\Omega$ which is roughly a 2-dimensional disk of radius R. One consequence of the boundary regularity theory mentioned in (A) is the following. There is a fixed $\epsilon > 0$, independent of R, with the property that whenever the boundary function ψ satisfies

$$\int_P |\nabla_T \psi|^2 \mathrm{d}A < \epsilon$$

then every minimizer φ is free of singularities in the region

$$K = \left\{ x : x \in \Omega, \operatorname{dist}(x, P) > \frac{1}{2} R\epsilon, \operatorname{dist}\left(x, P_{\frac{1}{2}}\right) < 2R\epsilon \right\},$$

here $P_{\frac{1}{2}}$ is the concentric disk of radius $\frac{1}{2}R$. Note that ϵ is dimensionless. Our **hot spot boundary regularity** theorem (proved in [AL]) asserts the existence of a fixed number $0 < \delta << \epsilon$ such that whenever $P' \subset P$ is a smaller subpatch of radius δR and

$$\int_{P \sim P'} |\nabla_T \psi|^2 \mathrm{d}A < \epsilon$$

then φ is also free of singularities in the region K above. In other words arbitrarly large boundary energy in a very small disk P' cannot by itself induce singularities far away.

3. COUNTING SINGULARITIES

The principal question motivating our work in [AL] is this:

How many singular points $N(\psi)$ is at possible for a minimizing φ to have?

The following possibilities seem plausible at the outset:

$N(\psi) \leq CE(\psi)$	FALSE;	
$N(\psi) \leq C \int_{\partial\Omega} \lvert J(\psi) \rvert \mathrm{d}A$	FALSE;	
$N(\psi) \leq C\partial\mathcal{E}(\psi)$	TRUE	**«The Linear Law»**.

here C is a constant, possibly depending on Ω.

The first possibility is false by counterexample – see below. The second possibility was suggested by the work in [BCL] and misled us for some time (had it been true it would have led to a beautiful geometric theory). In fact it is quite false as Example 1 below shows; in particular, $N(\psi)$ can be large while $J(\psi)$ vanishes identically.

Our main result, **The Linear Law**, is optimal (modulo the value of $C = C_\Omega$, of which we have no knowledge since our proof is by contradiction based on compactness arguments). It is, to our knowledge, the first bound of its kind.

The following example given by R. Hardt and F. H. Lin in [HL1] shows that $N(\psi)$ can indeed be proportional to $\partial\mathcal{E}(\psi)$. Choose N well separated small disks in $\partial\Omega$. Our ψ is constructed to wrap each disk D around the target sphere once (essentially by the inverse function to stereographic projection while preserving or reserving orientation as one chooses); each ∂D is mapped to the north pole. The complement of these disks in $\partial\Omega$ is mapped by ψ also to the north pole. Then $\partial\mathcal{E}(\psi) \approx CN$; the constant C is independent of the size of the disks since surface energy is scale invariant. Clearly the orientations of ψ on the disks can be arranged so that the total mapping degree of ψ is either zero or one. It is not hard to prove directly that any minimizing φ having ψ as boundary value function must have at least one singularity close to each tiny disk – otherwise $\mathcal{E}(\varphi)$ would be too large. Thus

$$N(\psi) \geq N \approx C^{-1}\partial\mathcal{E}(\psi).$$

Our first main new result (proved independently by Hardt and Lin in [HL2]) is that singularities cannot be very close if they are well inside Ω.

THEOREM 1. *There is a universal constant C (independent of Ω and ψ) such that whenever y and z in Ω are singular points of a minimizer φ then* $\operatorname{dist}(y,z) \geq C \operatorname{dist}(y,\partial\Omega)$.

The idea of the proof is the following. Fix y and suppose the contrary. Then there will be a sequence of minimizing $\varphi^{(j)}$ with singular points at $z^{(j)}$ and at

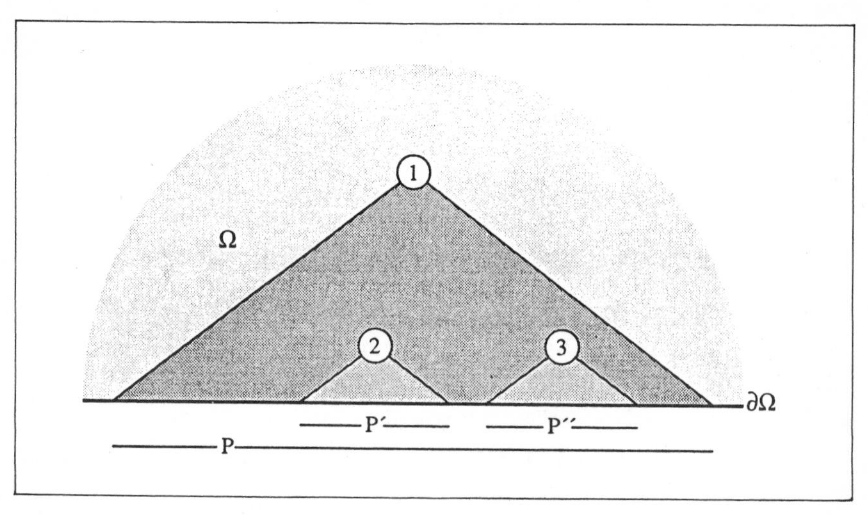

Fig. 3. Pictured here are the «cones of influence» in Ω of three singular points. The presence of singular points $1,2,3$ implies the presence of boundary energy in disks P, P', P'' in $\partial\Omega$. The problem is that these disks are not disjoint so that the total boundary energy is not a simple sum. Nesting of such cones induces a Cayley tree graph in which a combinatorial anaysis overcomes this difficulty.

y such that $z^{(j)} \to y$ as $j \to \infty$. A compactness argument (contradicting the negation) and monotonicity (A) shows that the energy of φ in small balls of radius R about y is *uniformly* greater than $8\pi R$. The limit of a subsequence of the minimizers $\varphi^{(j)}$ is a minimizer which thus can have at worst a singularity of degree ± 1 at y (by equation (2) above). The tangential approximation theorem implies that the energy of the limit φ must be very close to $8\pi R$ for a small R's. This leads to a contradiction because of the continuity of Dirichlet's integral when minimizers converge.

A consequence of Theorem 1 together with equation (2) above is the following.

THEOREM 2. **(Complete classification of energy minimizing maps from $\mathbf{R}^3$ to S^2.)** *Suppose $\varphi : \mathbf{R}^3 \to S^2$ is a minimizer. Then, either φ is a constant mapping or $\varphi = \pm\mathcal{R}\left(\frac{x-y}{|x-y|}\right)$ for some y and $\mathcal{R}$.*

Theorem 1 says that if there are many singularities they have to pile up near $\partial\Omega$. This leads to a difficult geometric-combinatorial problem on different scales proportional to δ^k, where δ is given in (E) above and $k = 1,2,,\ldots$ We attempt to illustrate this in Figure 3. Referring to the ϵ and δ of (E) consider the points 1, 2, and 3 in Ω at distances $R\epsilon$, $R\epsilon\delta$, and $R\epsilon\delta$ above a boundary patch P of radius R and two boundary patches P' and P'' of radii $R\delta$ inside P. The hot spot boundary regularity theorem gives us the following lower bounds for the energy of ψ in P if we consider the various possibilities of having singularities at positions 1, 2, or 3:

Positions occupied	**Local boundary energy**
(1 alone) or (2 alone) or (3 alone)	ϵ
(1 and 2) or (1 and 3) or (2 and 3)	2ϵ
(1 and 2 and 3)	2ϵ

The source of all our difficulties is that we cannot infer an energy 3ϵ if there are singularities at all three points.

If $S^{(k)}$ denotes the strip $\{x : x \in \Omega, \text{dist}(x, \partial\Omega) \leq \epsilon\delta^k\}$, we can effectively decompose each $S^{(k)}$ into cones of height $\epsilon\delta^k$ and base radius δ^k. We then have a Cayley tree whose vertices represent these cones (i.e. a vertex of order $k + 1$ is connected to a vertex of order k in the tree if the smaller cone is inside the larger one). A vertex is occupied if its cone has a singularity near the apex; otherwise it is unoccupied. Each occupied vertex gets an energy ϵ if and only if no more than one higher order vertex to which it is pathwise connected is occupied.

The actual details of decomposing each $S^{(k)}$ into cones so that due account is taken of overlaps (and all the other problems that will occur to the reader) involves a complicated covering and counting lemma. The final result is The Linear Law for $N(\psi)$ in terms of $\partial\mathcal{E}(\psi)$, as stated at the beginning of this section.

4. THREE EXAMPLES OF COUNTERINTUITIVE BEHAVIOR

EXAMPLE 1. **Zero Mapping Area.** It is easy to prove for any Ω that if ψ takes values only in some closed hemisphere of S^2 then φ has no singularities. We, however, are able to construct a single curve Γ in S^2 which is a slight perturbation of the equator and, for each N, a smooth boundary value function $\psi^N : \partial\Omega \to S^2$ **having its image equal to** Γ such than any minimizer φ^N having boundary values ψ^N must have at least N singular points. In the example of [AL], Ω is taken to be a ball, but the details of Ω are not important. The Jacobian $J(\psi^N)$ of each ψ^N vanishes identically since its image is one dimensional.

The idea behind the construction appears in the following preliminary problem. Consider reasonable mappings $\varphi : D^2 \to S^2$ from the unit disk D^2 in the plane having two dimensional Dirichlet's integral denoted by $\mathcal{E}_2(\varphi)$. Suppose $\Gamma \subset S^2$ is a smooth embedding of a circle parametrized by a map $P : \partial D^2 \to \Gamma$. The functions φ from D^2 to S^2 having boundary values P can be separated into two homological classes: the +class, in which, heuristically, φ «covers the top of S^2 one more time than it covers the bottom» and, the – class in which φ «covers the bottom one more time than it covers the top»; see Figure 4.

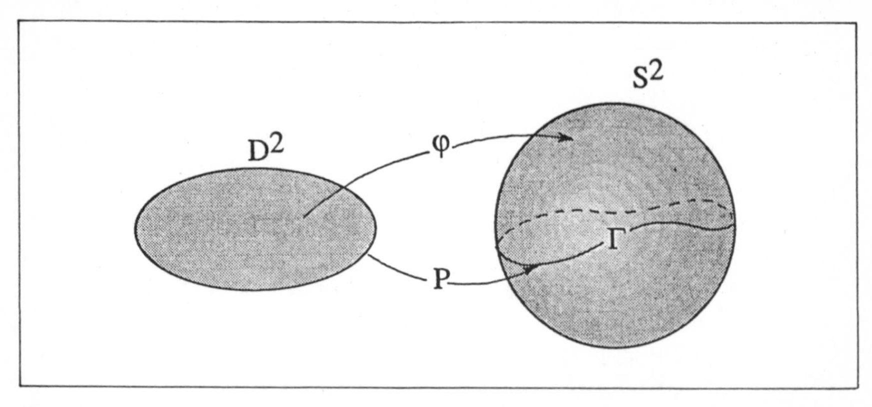

Fig. 4. Illustrated here is one of two homologically distinct classes of mappings $\varphi : D^2 \to S^2$ corresponding to a given boundary parametrization $P : \partial D^2 \to \Gamma$ (the curve Γ is a perturbation of the equator). A «+ function» is one which «covers the northern hemisphere». For some Γ's, the homology type preferred by a least energy mapping can change if the parametrization P is changed. This phenomenon leads ultimately to construction of least energy mappings from the ball to the sphere having many interior singularities but for which the boundary mapping of the sphere to the sphere has zero mapping area (its entire image lies within the curve Γ).

Consider the two numbers

$$E^{\pm}(P) = \inf\{\mathcal{E}_2(\varphi) : \varphi = P \text{ on } \partial D^2 \text{ and } \varphi \in \pm \text{ class}\}.$$

In general $E^+(P)$ will not be the same as $E^-(P)$.

We construct a single Γ having two different (homotopic) parametrizations P^+ and P^- such that

$$E^+(P^+) < E^-(P^+) - \epsilon \text{ and } E^-(P^-) < E^+(P^-) - \epsilon$$

for some $\epsilon > 0$. In other words if the parametrization of Γ changes from P^+ to P^- any *absolute* minimizer φ changes from lying in the + class to lying in the – class.

The next step is to let Ω be a very long solid tube T of radius 1 and length $N(L+1)$. (Actually, T is bent into a torus so that we can ignore the two ends.) As boundary function ψ we alternately paste P^- and P^+ on sections of length L (i.e. each cross-sectional disk has P^- or P^+ on its boundary). In the transitional regions of length 1 we smoothly interpolate between P^- and P^+ (which can be done since they are homotopic). In the transition regions ψ continues to take values only in Γ. See Figure 5.

If L is large enough (depending only on ϵ), it is believable (and we prove it) that φ must be mostly a – function on the P^- disks and it must be mostly a + function

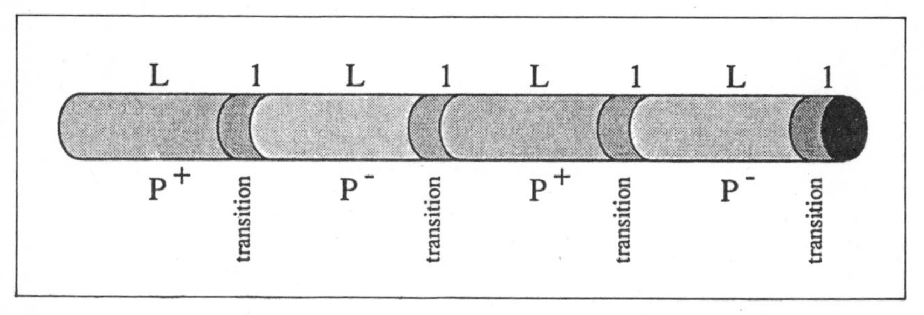

Fig. 5. Illustrated here is a boundary value function $\psi : \partial\Omega \to S^2$ for a long tube domain Ω. The image of ψ is a smooth curve Γ in S^2. On crossectional circles of $\partial\Omega$ the boundary values alternate between intervals of P^+ mappings and intervals of P^- separated by transition intervals. Least energy maps $\varphi : \Omega \to S^2$ with such boundary values map most crossections in P^+ regions to cover the northern hemisphere and map most crossections in P^- regions to cover the southern hemisphere. The minimizer φ therefore has at least one singular point near each transition region.

on the P^+ disks, for otherwise $\mathcal{E}(\varphi)$ would be unnecessarily large. But when φ switches from being a $-$ function to being a $+$ function φ must have a singularity for topological reasons. Thus, φ will have at least N singularities altogether.

The drawback to this example is that the domain T depends on N. To achieve the same result for a fixed domain Ω = unit ball, we first cut the surface ∂T longitudinally (i.e. perpendicular to the disks) and flatten it (key estimates here come from the conformal equivalence of the disk and the upper half plane and the fact that Dirichlet's integral in two dimensions is invariant under conformal reparametrizations of domains).This yields a strip of width 2π and length $N(L+1)$. We also rotate P^+ if necessary so that P^+ and P^- have the same value $\gamma \in \Gamma$ along the cut. Next we shrink the strip to width $(2\pi)^2/N(L+1)$ and length 2π. Finally we paste this strip (which is very narrow since N is large) along the equator of Ω and let $\psi : \partial\Omega \to S^2$ be the old ψ in the strip and let $\psi(x) = \gamma$ for $x \in \partial\Omega$ but $x \notin$ the strip. A somewhat nerve wracking argument shows, as expected, that any minimizer $\varphi : \Omega \to S^2$ must have at least N singularities close to the equator of Ω.

EXAMPLE 2. **Symmetry Breaking** When φ takes values in $\mathbb{R}^3$ instead of S^2, any geometric symmetry of Ω and ψ is inherited by the minimizing φ. The reason is simply that minimizers are unique in the linear case ($\Delta\varphi = 0$). When, as in our case, φ takes values in S^2, the symmetry of Ω and φ can be broken by φ; obviously there must then be several minimizers.

Let Ω be the unit ball in $\mathbb{R}^3$ and let $\psi : \partial\Omega \to S^2$ be the distortion of the identity map illustrated in Figure 6. In small caps N (resp. S) on $\partial\Omega, \psi$ covers

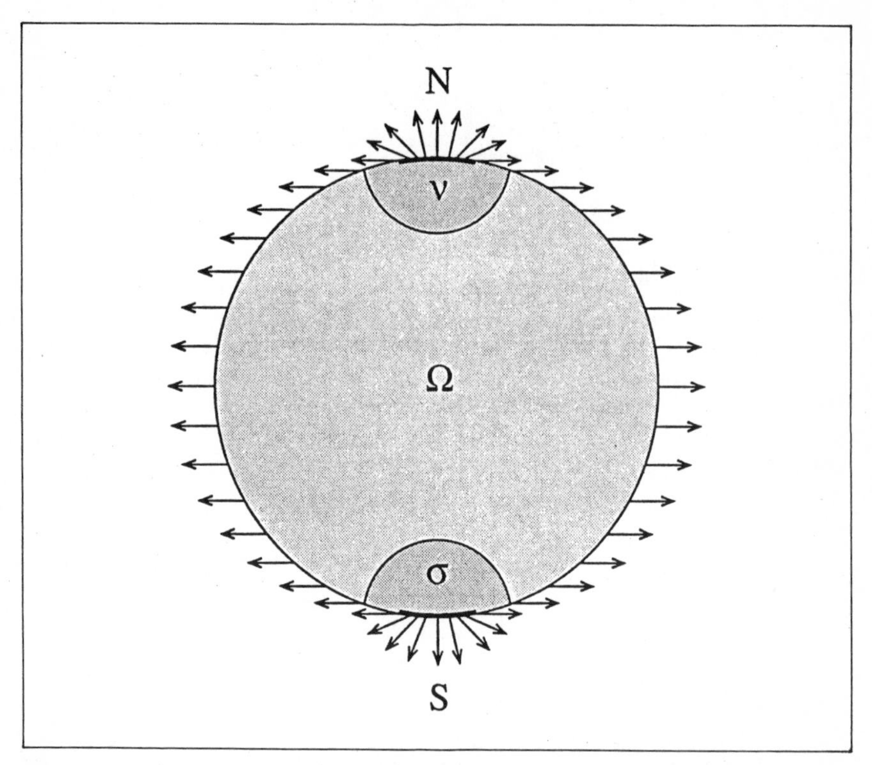

Fig. 6. Here our domain Ω is the unit ball so that $\partial\Omega$ is the unit sphere. Pictured schematically is a special boundary value function $\psi : \partial\Omega \to S^2$ having a mirror image symmetry through the equatorial plane. A small cap N around the north pole maps to cover the entire northern hemisphere of S^2 while a small cap S around the south pole covers the entire southern hemisphere. The sphere less these two caps maps entirely to the equator. Longitude is preserved in each of these regions. No minimizing $\varphi : \Omega \to S^2$ having boundary values ψ can possess such a symmetry since the (necessarily odd) number of singular points must be contained within one of the regions ν and σ near the poles.

the northern (resp. southern) hemisphere of S^2. The two maps are mirror images of each other. On the rest of $\partial\Omega$ between N and S, ψ takes values in the equator of S^2 in the obvious way, i.e. $\psi(x, y, z) = (x^2 + y^2)^{-1/2}(x, y, 0)$.

THEOREM 3. *Any minimizer φ can have singularities only in small shaded regions in Ω, labelled ν and σ, near the caps N and S.*

Since $\deg(\psi) = 1$, this result implies that φ does not inherit the mirror image symmetry through the equatorial disk possessed by ψ. (Our function φ necessarily has an odd number of singularities, and if φ were symmetric, it would necessarily have one on the equatorial disk in Ω.)

The proof of Theorem 3 has two parts. First we show that when N and S are small φ has no singularities in a concentric ball Ω' of radius $1-\epsilon$ for some small ϵ. This is done by a variational (or comparison) argument. Second, we show that there are no singularities in $\{x : 1 \geq |x| > 1-\epsilon \text{ and } \operatorname{dist}(x, \sigma \cap \nu) > \epsilon\}$ by using the boundary regularity (E).

EXAMPLE 3. **Boiling Water** The [BCL] result mentioned in (D) above suggests that + and − singularities tend to annihilate each other. On the other hand, the hot spot boundary regularity mentioned in (E) above suggests that behavior at different length scales (as measured by the distance to $\partial\Omega$) is independent so that + and − singularities *could* coexist provided their distances to $\partial\Omega$ were very different. There would appear to be a conflict here and one of our results is that of the two points of view just mentioned the second one is correct. We have proved the following.

THEOREM 4. *Let Ω be the unit ball and let $p_1, \ldots, p_M$ be any distinct points in $\partial\Omega$. Also let $N_1, \ldots, N_M$ be any positive integers and for each $i = 1, \ldots, M$ let A_i be any sequence of length N_i consisting of $+1$'s and -1 's. Finally, let $\epsilon > 0$. Then there is a smooth $\psi : \partial\Omega \to S^2$ such that*

(i) $\partial\mathcal{E}(\psi) \leq \epsilon + 8\pi \sum_{i=1}^{M} N_i$.

(ii) The minimizer φ is unique.

(iii) For each $i = 1, \ldots, M$ there are at least N_i singularities stacked nearly vertically above p_i (like bubbles in a pan of water that is about to boil), and these have the specified sequence of degrees given by A_i.

REFERENCES

[ABL] F. ALMGREN, W. BROWDER and E. LIEB: Co-area, liquid crystals and minimal surfaces. In: *Partial Differential Equations,* ed. S. S. Chern, Springer Lecture Notes in Math., **1306**, 1-12 (1988).

[AL] F. ALMGREN and E. LIEB: Singularites of energy minimizing maps from the ball to the sphere: examples counterexamples and bounds. *Ann. of Math.*, **128**, 483-530 (1988). See also: Singularities of energy minimizing maps from the ball to the sphere, *Bull. Amer. Math. Soc.*, **17**, 304-306 (1987).

[BCL] H. BREZIS, J-M. CORON and E. LIEB: Harmonic maps with defects. *Commun. Math. Phys.* **107**, 649-705 (1986).

[HL1] R. HARDT and F. H. LIN: A remark on H^1 mappings. *Manuscripta Math.*, **56**, 1-10 (1986).

[HL2] R. HARDT and F. H. LIN: Stability of singularities of minimizing harmonic maps. *J. Diff. Geom.*, **29**, 113-123 (1989).

[K] M. KLÉMAN: Points, lignes, parois dans les fluides anisotropes et les solides cristalline. *Les Éditiones de Physique (Orsay)*, **I**, 36-37.

[S] L. SIMON: Asymptotics for a class of nonlinear evolution equations with applications to geometric problems. *Ann. of Math.* **118**, 525-571 (1983).

[SU1] R. SCHOEN and K. UHLENBECK: A regularity theory for harmonic maps. *J. Diff. Geom.*, **17**, 307-335 (1982).

[SU2] R. SCHOEN and K. UHLENBECK: Boundary regularity and the Dirichlet problem of harmonic maps. *J. Diff. Geom.*, **18**, 253-268 (1983).

ON THE CONFORMAL CAPACITY PROBLEM

G. A. PHILIPPIN[1] - L. E. PAYNE[2]

1. INTRODUCTION

In this lecture we present a variety of results for the following boundary value problem defined in a noncontractable region Ω of $\mathbb{R}^N$, sufficiently regular, bounded internally by Γ_1 and externally by Γ_0 :

$$(g(q^2)u_{,i})_{,i} = 0 \quad \text{in} \quad \Omega, \tag{1.1}$$

$$u = 0 \quad \text{on} \quad \Gamma_0, u = 1 \quad \text{on} \quad \Gamma_1. \tag{1.2}$$

In (1.1), g is a given C^1 function of its argument, and q^2 stands for $|\text{grad}u|^2$. A subscript i preceded by a comma denotes partial differentiation with respect to $x_i, i = 1,\ldots,N,$ and the summation convention over repeated indices is used throughout. The differential equation (1.1) will usually be assumed to be elliptic. This is the case for nonnegative g satisfying the condition

$$G(s) := g(s) + 2sg'(s) > 0. \tag{1.3}$$

If we select $g \equiv 1$, (1.1), (1.2) may be regarded as the electrostatic capacity problem of two conductors. If we select $g \equiv q^{N-2}$, the corresponding B.V.P. (1.1), (1.2) is refered to as the conformal capacity problem, and the equation (1.1) will be uniformly elliptic provided q does not vanish in $\bar{\Omega}$. This problem plays a useful role in the theory of quasi conformal mappings. The interested reader is refered to a paper of G. Anderson [1] and to the bibliography cited there.

[1] Research supported by the Natural Sciences and Engineering Research Council Canada.
[2] Research supported in part by NSF Grant # DMS-8600250.

In the second section of this paper, we establish a maximum principle for a particular combination of u and $|\text{grad}u|$, where u is a classical solution of (1.1). This maximum principle is an extension of previous results obtained by the authors for harmonic functions in [10, 11]. We refer to the book of Sperb [13] and to a survey paper of Payne [9] for more general information.

Applications of this maximum principle for the conformal capacity problem are then presented in the third section. In particular we establish various isoperimetric bounds for $u, |\nabla u|$, the conformal capacity defined in (3.4), and some geometric quantities associated with Ω. We show further that if we add Bernoulli type boundary conditions (i.e. $|\text{grad}u| = \text{const.}$ on the two boundaries assumed to be starshaped), then the overdetermined conformal capacity problem is solvable only if the two boundaries are concentric spheres. In Section 4, we establish an N-dimensional version of Longinetti's results [8]. For instance we establish some convexity properties for the surface area of the level sets of u under appropriate convexity assumptions on the two surfaces Γ_0 and Γ_1.

2. THE MAXIMUM PRINCIPLE

The main result of this section is stated in the next

THEOREM 1. *The function $\phi(x)$ defined on solutions of (1.1) as*

$$\phi(x) := f(u)\int_0^{q^2} G(s)\,\mathrm{d}s, \tag{2.1}$$

with $G(s) := g(s) + 2sg'(s)$, satisfies the following second order elliptic differential inequality

$$\begin{aligned} g\Delta\phi + 2g'\phi_{,k\ell}u_{,k}u_{,\ell} + W_k\phi_{,k} \geq q^2 G(q^2)\int_0^{q^2} G(s)\,\mathrm{d}s \\ \left\{f''(u) - \frac{f'^2(u)}{f(u)}\left[2 - \frac{N}{2(N-1)q^2 g(q^2)}\int_0^{q^2} G(s)\,\mathrm{d}s\right]\right\} \end{aligned} \tag{2.2}$$

in Ω, where W_k is a vector field regular throughout Ω, except at critical points of u. Here, f is assumed to be a positive function.

For the proof, we need the following

LEMMA 1. *On solutions of (1.1) we have the inequality*

$$u_{,ij}u_{,ij} \geq \frac{u_{,ij}u_{,j}u_{,ik}u_{,k}}{q^2} + \frac{G^2(q^2)}{(N-1)g^2(q^2)}\frac{(u_{,ik}u_{,i}u_{,k})^2}{q^4}, \tag{2.3}$$

valid at all noncritical points of u $(q > 0)$.

The proof of Lemma 1 is based on the fact that for any χ_{ij},

$$\chi_{ij}\chi_{ij} \geq 0. \tag{2.4}$$

Choosing

$$\chi_{ij} := u_{,ij} - \frac{u_{,ik}u_{,k}u_{,j}}{q^2} + \frac{u_{,\ell k}u_{,\ell}u_{,k}}{q^2}\frac{G(q^2)}{(N-1)g(q^2)}\left\{\delta_{ij} - \frac{u_{,i}u_{,j}}{q^2}\right\} \tag{2.5}$$

in (2.4) (where δ_{ij} is the Kronecker symbol) and solving with respect to $u_{,ij}u_{,ij}$ we are led to the desired result.

For the proof of Theorem 1, we compute first and second order derivatives of ϕ:

$$\phi_{,k} = f_{,k}\int_0^{q^2} G(s)\,\mathrm{d}s + fG(q^2)(q^2)_{,k}, \tag{2.6}$$

$$\begin{aligned}\phi_{,k\ell} = f_{,k\ell}\int_0^{q^2} G(s)\,\mathrm{d}s &+ G(q^2)[f_{,k}(q^2)_{,\ell} + f_{,\ell}(q^2)_{,k}] \\ &+ fG'(q^2)_{,k}(q^2)_{,\ell} + fG(q^2)_{,k\ell},\end{aligned} \tag{2.7}$$

$$\Delta\phi = \Delta f\int_0^{q^2} G(s)\,\mathrm{d}s + 2Gf_{,k}(q^2)_{,k} + fG'(q^2)_{,k}(q^2)_{,k} + fG\Delta(q^2). \tag{2.8}$$

The last term in (2.8) may of course be written as follows:

$$\Delta(q^2) = 2u_{,j}\Delta u_{,j} + 2u_{,ij}u_{,ij}. \tag{2.9}$$

The first expression on the right hand side of (2.9) may be computed from the differential equation (1.1) rewritten as

$$\Delta u = -(\log g)'(q^2)_{,i}u_{,i}. \tag{2.10}$$

Differentiating (2.10), we obtain

$$\begin{aligned}u_{,j}\Delta u_{,j} = &-(\log g)''[(q^2)_{,i}u_{,i}]^2 - (\log g)'(q^2)_{,ij}u_{,i}u_{,j} \\ &- \frac{1}{2}(\log g)'(q^2)_{,i}(q^2)_{,i}.\end{aligned} \tag{2.11}$$

Using (2.7), (2.8), (2.9), (2.10) and (2.3), we arrive at the inequality

$$
\begin{aligned}
\Delta\phi + 2(\log g)'\phi_{,k\ell}u_{,k}u_{,\ell} \geq \int_0^{q^2} G(s)\,ds[\Delta f + 2(\log g)' f_{,k\ell}u_{,k}u_{,\ell}] + \\
+ 2Gf_{,k}(q^2)_{,k} + f(q^2)_{,k}(q^2)_{,k}\left\{G' - G(\log g)' + \frac{G}{2q^2}\right\} + \\
+ 2f[(q^2)_{,k}u_{,k}]^2 \quad [-G(\log g)'' + G'(\log g)'] + \\
+ \frac{G^3 f}{2(N-1)q^4 g^2}[(q^2)_{,k}u_{,k}]^2 + 4(\log g)' G f_{,i}u_{,i}(q^2)_{,k}u_{,k}.
\end{aligned}
\tag{2.12}
$$

The next step is to rewrite those terms in (2.12) containing $(q^2)_{,k}(q^2)_{,k}, (q^2)_{,k}u_{,k}$, and $(q^2)_{,k}f_{,k}$ using successively

$$
\phi_{,k}(q^2)_{,k} = f_{,k}(q^2)_{,k}\int_0^{q^2} G(s)\,ds + fG(q^2)_{,k}(q^2)_{,k},
\tag{2.13}
$$

$$
\phi_{,k}u_{,k} = f_{,k}u_{,k}\int_0^{q^2} G(s)\,ds + fG(q^2)_{,k}u_{,k},
\tag{2.14}
$$

$$
\phi_{,k}f_{,k} = f_{,k}f_{,k}\int_0^{q^2} G(s)\,ds + fG(q^2)_{,k}f_{,k}.
\tag{2.15}
$$

We are then led to

(2.16)

$$
\begin{aligned}
L\phi := \Delta\phi + 2(\log g)'\phi_{,k\ell}u_{,k}u_{,\ell} + W_k\phi_{,k} \geq \\
\geq \int_0^{q^2} G\,ds[\Delta f + 2(\log g)' f_{,k\ell}u_{,k}u_{,\ell}] + \\
+ \frac{f_{,k}f_{,k}}{f}\int_0^{q^2} G\,ds\left\{-2 + G^{-2}\int_0^{q^2} G\,ds\right. \\
\left.\left[G' - G(\log g)' + \frac{G}{2q^2}\right]\right\} + \\
+ \frac{(u_{,k}f_{,k})^2}{f}\int_0^{q^2} G\,ds\left\{2G'G^{-2}(\log g)'\int_0^{q^2} G\,ds - \right. \\
\left. -2G^{-1}(\log g)''\int_0^{q^2} G\,ds + \frac{G}{2(N-1)g^2q^4}\int_0^{q^2} G\,ds - 4(\log g)'\right\},
\end{aligned}
$$

for some vector field W_k, regular throughout Ω, except at possible critical points of u. We compute now

$$\frac{f_{,k}f_{,k}}{f} = \frac{f'^2}{f}q^2, \tag{2.17}$$

$$\frac{f_{,k}u_{,k}}{f} = \frac{f'q^2}{f}, \tag{2.18}$$

and

$$\Delta f + 2(\log g)' f_{,k\ell}u_{,k}u_{,\ell} = \frac{q^2 G}{g}f''. \tag{2.19}$$

The insertion of (2.17), (2.18), and (2.19) into (2.16) leads to the desired result after some reduction.

As a consequence of the differential inequality established in Theorem 1, we have the following.

THEOREM 2. *If we select a positive C^2 function f which satisfies the condition*

$$f''(u) - \frac{f'^2(u)}{f(u)}\left\{2 - \frac{N}{2(N-1)q^2 g(q^2)}\int_0^{q^2} G(s)\,\mathrm{d}s\right\} \geq 0 \tag{2.20}$$

on solutions of (1.1), then the function $\phi(x)$ defined as

$$\phi(x) := f(u)\int_0^{q^2} G(s)\,\mathrm{d}s, \tag{2.21}$$

takes its maximum value at some point P on the boundary $\partial\Omega$ of Ω.

The proof of Theorem 2 follows from the constructed elliptic differential inequality $L\phi \geq 0$ in Ω, together with Hopf's first principle [4, 12], and the observation that

$$\phi \geq 0 \quad \text{in} \quad \Omega, \tag{2.22}$$

with equality at critical points of u, so that critical points are obviously not candidate for points of maximum value of ϕ. We conclude therefore that ϕ takes its maximum value on $\partial\Omega$.

We note that condition (2.20) involves not only $f(u)$, but a function of q^2, so that an upper bound for the quantity $q^{-2} g^{-1} \int_0^{q^2} G(s)\,\mathrm{d}s$ is needed, if we are to obtain an admissible function f from (2.20). The easiest case in this respect is

$$g(s) := s^{p/2}, G(s) = (1+p)\,s^{p/2}. \tag{2.23}$$

In this case we have

$$\frac{1}{q^2 g(q^2)} \int_0^{q^2} G(s)\,\mathrm{d}s = \frac{2(1+p)}{p+2} = \text{const}., \tag{2.24}$$

and condition (2.20) may be rewritten as

$$f'' - \left(2 - \frac{N(1+p)}{(N-1)(p+2)}\right) \frac{f'^2}{f} \geq 0, \tag{2.25}$$

or

$$\begin{cases} (p+2-N) \left\{ f^{\frac{p+2-N}{(N-1)(p+2)}} \right\}'' \geq 0 & \text{if } p \neq N-2, \\ (\log f)'' \geq 0 & \text{if } p = N-2. \end{cases} \tag{2.26}$$

The particular case of a harmonic function u corresponding to $p := 0$ has already been investigated by the authors in [10, 11]. We wish now to investigate the case of the conformal capacity problem corresponding to $p := N-2$. For this problem, (2.25) and (2.26) become merely

$$f'' - \frac{f'^2}{f} = (\log f)'' \geq 0, \tag{2.27}$$

and we conclude from Theorem 2 that the function

$$\phi(x) := e^{\alpha u} |\text{grad} u|^N \tag{2.28}$$

takes its maximum on $\partial\Omega$ for any $\alpha \in \mathbb{R}$. This result will be exploited in Section 3. Note that with the g of (2.23), L, defined in (2.16), will be uniformly elliptic outside compact neighborhoods of critical points of u.

3. ISOPERIMETRIC BOUNDS FOR u AND $|\nabla u| = q$ IN THE CONFORMAL CAPACITY PROBLEM

We now restrict our attention to the conformal capacity problem

$$(|\nabla u|^{N-2} u_{,i})_{,i} = 0 \quad \text{in} \quad \Omega \subset \mathbb{R}^N, \tag{3.1}$$

$$u = 0 \quad \text{on} \quad \Gamma_0, u = 1 \quad \text{on} \quad \Gamma_1, \tag{3.2}$$

where the exterior boundary Γ_0 and the interior boundary Γ_1 are assumed to have at each point a well defined exterior unit normal vector $\vec{n} := (n_1, \ldots, n_N)$. Then from Hopf's second principle [5, 12] we have

$$q = -\frac{\partial u}{\partial n} > 0 \quad \text{on} \quad \Gamma_i, i = 0, 1. \tag{3.3}$$

The quantity of interest is the conformal capacity C defined as

$$C := \int_\Omega q^N \mathrm{d}x = \oint_{\Gamma_u} \left| \frac{\partial u}{\partial n} \right|^{N-1} \mathrm{d}s, \tag{3.4}$$

where Γ_u is an arbitrary level set of u. We have used (2.1) and integration by parts to establish the last equality.

Let us now select $\alpha \in \mathbb{R}$ in such a way that the function ϕ defined in (2.28) takes its maximum value on each boundary $\Gamma_i, i = 0, 1$. Using the notation

$$q_i := \max_{\Gamma_i} q, \quad i = 0, 1, \tag{3.5}$$

α has to be chosen so that

$$q_1^N e^\alpha = q_0^N. \tag{3.6}$$

This implies that

$$\alpha = N \log \frac{q_0}{q_1}. \tag{3.7}$$

From this point on we shall use this particular choice for α.

From Theorem 2, we have

$$\phi := e^{\alpha u} q^N \leq q_0^N, \tag{3.8}$$

or

$$q \le q_0^{1-u} q_1^u \quad \text{in} \quad \Omega . \tag{3.9}$$

This shows in particular that the function $Q(u) := \max_{\Gamma_u} q$ is log-convex.

We have equality in (3.8) and in (3.9) if and only if $\phi \equiv$ const . throughtout Ω. This implies in particular that $u(x)$ is radially symmetric. Endeed, if $\phi \equiv$ const . throughout $\bar{\Omega}$, then we must have equality in (2.2). This fact implies further that χ_{ij} defined in (2.5) vanish identically in Ω :

$$\chi_{ij} = u_{,ij} - \frac{u_{,ik}u_{,k}u_{,j}}{q^2} + \frac{u_{,\ell k}u_{,\ell}u_{,k}}{q^2}\left\{\delta_{ij} - \frac{u_{,i}u_{,j}}{q^2}\right\} \equiv 0 \quad \text{in} \quad \Omega . \tag{3.10}$$

With $\phi_{,k} \equiv 0$ in Ω and $q > 0$ in $\bar{\Omega}$, (3.10) reduces to

$$\left(\frac{u_{,i}}{q^2}\right)_{,j} = \frac{\alpha}{N}\,\delta_{ij} \quad \text{in} \quad \Omega , \tag{3.11}$$

where α is given in (3.7). A first integration of (3.11) leads to

$$\frac{u_{,i}}{q^2} = \frac{\alpha}{N} x_i , i = 1, \ldots, N, \tag{3.12}$$

for a suitable choice of the origin, from which we obtain

$$u = \frac{\log \frac{r}{R_0}}{\log \frac{q_0}{q_1}} \quad \text{in} \quad \Omega , \tag{3.13}$$

where $r := \sqrt{x_k x_k}$ measures the distance from the origin, and R_0 is a positive constant. This result shows in particular that Γ_0 and Γ_1 are concentric N-spheres whose radii R_0, R_1 are related as follows.

$$R_0 q_0 = R_1 q_1 . \tag{3.14}$$

Let us now compute the normal derivative of ϕ at points $P_i \in \Gamma_i, i = 0, 1,$ at which ϕ assumes its maximum value. We must have

$$\frac{\partial \phi}{\partial n} = -e^{\alpha u} q^{N-1} \quad (\alpha q^2 + N u_{,nn}) \begin{cases} \ge 0 & \text{at } P_0 \in \Gamma_0 , \\ \le 0 & \text{at } P_1 \in \Gamma_1 . \end{cases} \tag{3.15}$$

Using the differential equation (3.1) rewritten in normal coordinates as

$$u_{nn} + K u_n = u_{nn} - Kq = 0 \quad \text{on} \quad \partial\Omega, \tag{3.16}$$

we obtain from (3.15) the following bounds for the average curvatures $K_i(P_i)$ of Γ_i at $P_i, i = 0, 1$:

$$K_0(P_0) \leq -\frac{\alpha q_0}{N} = -q_0 \log \frac{q_0}{q_1}, \tag{3.17}$$

$$K_1(P_1) \geq -\frac{\alpha q_1}{N} = -q_1 \log \frac{q_0}{q_1}. \tag{3.18}$$

We are now in the position to establish the following result:

THEOREM 3. *If Γ_0 and Γ_1 are starshaped, then the B.V.P. (3.1), (3.2), complemented to an overdetermined system by imposing the further Bernoulli conditions*

$$q = q_i = \text{const}. \quad \text{on} \quad \Gamma_i, i = 0, 1, \tag{3.19}$$

is solvable if and only if Γ_0 and Γ_1 are concentric N-spheres whose radii R_0, R_1 are related by (3.14).

For the proof of Theorem 3, we observe that ϕ takes its maximum value at each point of $\partial\Omega = \Gamma_0 \cup \Gamma_1$ under condition (3.19), so that (3.17) remains true at every point P_0 of Γ_0, and (3.18) at every point P_1 of Γ_1. Since Γ_0 and Γ_1 are starshaped, we may use (3.17) and (3.18) to derive the following bounds for the measures S_i of $\Gamma_i, i = 0, 1$:

$$S_0 = \oint_{\Gamma_0} h_0 K_0 \, \mathrm{d}s \leq \max_{\Gamma_0} K_0 \oint_{\Gamma_0} h_0 \, \mathrm{d}s = N V_0 K_{0_{\max}} \leq -N V_0 q_0 \log \frac{q_0}{q_1}, \tag{3.20}$$

$$S_1 = \oint_{\Gamma_1} h_1 K_1 \, \mathrm{d}s \geq \min_{\Gamma_1} K_1 \oint_{\Gamma_1} h_1 \, \mathrm{d}s = N V_1 K_{1_{\min}} \geq -N V_1 q_1 \log \frac{q_0}{q_1}, \tag{3.21}$$

where h_i are the support functions associated with Γ_i, and V_i are the N-volumes of the domains enclosed by $\Gamma_i, i = 0, 1$.

From (3.20) and (3.21), we obtain

$$\frac{S_0}{S_1} \leq \frac{V_0}{V_1} \cdot \frac{q_0}{q_1} = \left(\frac{q_1}{q_0}\right)^{N-1} = \frac{S_0}{S_1}, \tag{3.22}$$

where the middle equality is a consequence of Rellich's identity

$$NV_0 q_0^N = \oint_{\Gamma_0} \left|\frac{\partial u}{\partial n}\right|^N h_0 \, \mathrm{d}s = \oint_{\Gamma_1} \left|\frac{\partial u}{\partial n}\right|^N h_1 \, \mathrm{d}s = NV_1 q_1^N, \tag{3.23}$$

and the last equality in (3.22) is a consequence of (3.4). It follows from (3.22) that we must have equality in both (3.17) and (3.18). We then conclude, using Hopf's second principle [5.12], that $\phi \equiv$ const . throughout Ω. This fact leads to the desired result as already shown.

From the above analysis, we conclude that the inequalities (3.8) , (3.9), (3.17) and (3.18) are strict unless the two boundaries are concentric N-spheres, in which case $\phi(x)$ is constant.

Suppose now that the average curvature K_0 of the exterior boundary Γ_0 is nonnegative. In this case we have

$$\frac{\partial}{\partial n}(q^2) = 2\, u_n u_{nn} = -2\, K_0 q^2 \leq 0 \tag{3.24}$$

on Γ_0. This means that q takes its maximum value on the interior boundary Γ_1, so that we have

$$q_{\max} = q_1 > q_0 \,. \tag{3.25}$$

Let us integrate (3.8) rewritten as

$$e^{\frac{\alpha u}{N}} \mathrm{d}u \leq q_0 \, \mathrm{d}r \tag{3.26}$$

along a straight line segment joining an arbitrary point on Γ_0 to some point P in Ω. We then obtain the inequality

$$0 \leq 1 - \left(\frac{q_0}{q_1}\right)^u \leq -q_0 \log \frac{q_0}{q_1} \delta_0 \leq \frac{q_0}{q_1} K_{1_{\max}} \delta_0 , \tag{3.27}$$

where we have used (3.18). Here δ_0 is the distance from Γ_0 to P. If we integrate (3.26) along a straight line segment joining $P \in \Omega$ to an arbitrary point of Γ_1, we obtain

$$0 \leq \left(\frac{q_0}{q_1}\right)^u - \frac{q_0}{q_1} \leq -q_0 \log \left(\frac{q_0}{q_1}\right) \delta_1 \leq \frac{q_0}{q_1} K_{1_{\max}} \delta_1 . \tag{3.28}$$

Here δ_1 is the distance from P in Ω to the boundary Γ_1. As particular cases of (3.27) or (3.28), we obtain

$$0 \leq q_1 - q_0 \leq -q_0 q_1 \log \left(\frac{q_0}{q_1}\right) \delta \leq q_0 K_{1_{\max}} \delta, \tag{3.29}$$

where δ is the distance between the two boundaries Γ_0 and Γ_1.

The basic inequality (3.8) (as well as subsequent inequalities) may be used in various ways to construct further isoperimetric inequalities involving e.g. the volume V of Ω (enclosed by the two boundaries), the conformal capacity C of Ω, and the geometric quantities already considered. This will be illustrated in the following two examples that lead to the isoperimetric inequalities (3.33) and (3.37) below. Let us consider the volume $V(\bar{u})$ of the subset $\Omega_{\bar{u}} := \{x \in \Omega \,|\, 0 < u(x) < \bar{u}\}$. We have

$$V(\bar{u}) = \int_0^{\bar{u}} \oint_{\Gamma_\eta} \frac{\mathrm{d}\,s}{|\mathrm{grad} u|} \mathrm{d}\,\eta, \tag{3.30}$$

where $\Gamma_\eta := \{x \in \Omega \,|\, u(x) = \eta\}$ are the level sets of $u, 0 \leq \eta \leq 1$. Using (3.8), the expression $|\mathrm{grad} u|^{-1}$ in (3.30) may be estimated as follows

$$|\mathrm{grad} u|^{-1} \geq |\mathrm{grad} u|^{N-1} q_0^{-N} e^{-\alpha u}. \tag{3.31}$$

From (3.30), (3.31), and (3.4), we obtain

$$V := V(1) \geq \frac{C[q_0^{-N} - q_1^{-N}]}{N \log \frac{q_1}{q_0}} \geq \frac{C[q_0^{-N} - q_1^{-N}]}{N \log(1 + K_{1_{\max}} \delta)}, \tag{3.32}$$

where we have used (3.29) for the last step. The above inequality may be rewritten as

$$0 \leq q_0^{-N} - q_1^{-N} \leq N C^{-1} V \log(1 + K_{1_{\max}} \delta). \tag{3.33}$$

Also from (3.26) we have, applying the Schwarz inequality and using (3.8),

$$\frac{\partial V}{\partial u} = \oint_{\Gamma_u} \frac{\mathrm{d}\,s}{|\mathrm{grad} u|} \geq \frac{S^2(u)}{\oint_{\Gamma_u} |\mathrm{grad} u| \mathrm{d}\,s} \geq q_0^{-1} S(u) e^{\frac{\alpha u}{N}}, \tag{3.34}$$

where $S(u)$ is the measure of Γ_u. $S(u)$ may be estimated from below in terms of $V_0 - V(u)$ by means of the classical isoperimetric geometric inequality

$$S(u) \geq N^{\frac{N-1}{N}} \omega_N^{1/N} [V_0 - V(u)]^{\frac{N-1}{N}}, \tag{3.35}$$

where $\omega_N := \frac{2\pi^{N/2}}{\Gamma(N/2)}$ is the surface area of the unit sphere in $\mathbf{R}^N$. If we combine (3.34) and (3.35), we obtain a differential inequality for $V(u)$. Integrating this inequality, we obtain

$$V_0^{1/N} - V_1^{1/N} \geq \left(\frac{\omega_N}{N}\right)^{1/N} \frac{[q_0^{-1} - q_1^{-1}]}{\log \frac{q_1}{q_0}} \geq \left(\frac{\omega_N}{N}\right)^{1/N} \frac{q_1 - q_0}{q_0 K_{1_{\max}}}, \tag{3.36}$$

where we have used (3.18) for the last step, or

$$0 \leq q_1 - q_0 \leq \left(\frac{N}{\omega_N}\right)^{1/N} q_0 K_{1_{\max}} \left[V_0^{1/N} - V_1^{1/N}\right]. \tag{3.37}$$

We recall that V_i are the volume of the domains enclosed by $\Gamma_i, i = 0, 1$.

4. CONVEXITY PROPERTIES OF $S(u)$ AND $V(u)$

This section is almost independent of the beginning of the paper, except for the notation. We consider the general boundary value problem (1.1), (1.2) in a convex noncontractable region $\Omega \subset \mathbf{R}^N$, i.e. Γ_0 and Γ_1 enclose two convex domains. We assume moreover that the level sets Γ_u enclose convex domains. This is known to be true in many particular cases including (2.23), as demonstrated by Lewis in [7] and by other authors [2, 3, 6]. Moreover there is no critical point of u in $\bar{\Omega}$ under the above convexity assumption. The main purpose of this section is to establish an N-dimensional version of Longinetti's convexity property of the measure $S(u)$ of the level sets Γ_u [8]. The result may be stated as follows:

THEOREM 4. *Under the above convexity assumption of the level sets of u, we have*

$$S''(u)S(u) - \mathcal{E}S'^2(u) \geq 0, \tag{4.1}$$

with

$$\mathcal{E} := \min_{\Omega} \frac{g(q^2)}{G(q^2)} > 0. \tag{4.2}$$

The proof of Theorem 4 is based on a straightforward computation of $S'(u)$ and $S''(u)$, and makes use of the following differential geometric result:

LEMMA 2. *Let $\vec{n} := (n_1, \ldots, n_N)$ be the unit normal vector associated with an oriented surface Γ in $\mathbf{R}^N, N \geq 2$. Then the following identity holds:*

$$n_{i,j}n_{i,j} = (N-1)^2 K^2 - 2(N-2)J + n_{i,k}n_k n_{i,\ell}n_\ell, \tag{4.3}$$

with

$$K := \frac{1}{N-1}\sum_{i=1}^{N-1} k_i, \tag{4.4}$$

(4.5) $$J := \frac{1}{N-2} \sum_{i,j=1}^{N-1} k_i k_j \text{ (double summation over } j > i).$$

In (4.4) and (4.5), $k_i, i = 1, \ldots, N-1$ *are the* $N-1$ *principal curvatures of* Γ. K *defined in (4.4) is the average curvature of* Γ *already introduced in the previous sections.* J *given in (4.5) is defined only for* $N \geq 3$. *If* $N = 3$, *it represents the gaussian curvature of* Γ. *If* $N = 2$, J *is absent from (4.3).*

For the proof of Lemma 2, we consider the curvature matrix

(4.6) $$B_{ij} := n_{i,j} - n_{i,k} n_k n_j$$

associated with the surface Γ. Since Γ may be represented locally as $u(x) = 0$, we have $n_i = \frac{-u_{,i}}{\sqrt{u_{,k}u_{,k}}}$, so that the matrix B_{ij} is symmetric. Moreover we have $B_{ij} n_j = 0$, i.e. n_j is an eigenvector of the curvature matrix B_{ij} whose corresponding eigenvalue is zero. The $N-1$ other real eigenvalues of B_{ij} are precisely the principal curvatures k_i of Γ, $i = 1, \ldots, N-1$. It follows then that

(4.7) $$B_{ij} B_{ji} = n_{i,j} n_{i,j} - n_{i,j} n_j n_{i,\ell} n_\ell = \sum_{i=1}^{N-1} k_i^2 = (N-1)^2 K^2 - 2(N-2) J,$$

where K and J are defined in (4.4) and in (4.5).

We compute now the first and second order derivatives of $S(u)$.

We have

(4.8)
$$\begin{aligned} S_0 - S(u) &= \oint_{\partial\Omega_u} \mathrm{d}s = \oint_{\partial\Omega_u} n_i n_i \mathrm{d}s = \\ &= \int_{\Omega_u} n_{i,i} \mathrm{d}x = (N-1) \int_{\Omega_u} K \mathrm{d}x = \\ &= (N-1) \int_0^u \oint_{\Gamma_\eta} \frac{K \mathrm{d}s}{|\mathrm{grad} u|} \mathrm{d}\eta. \end{aligned}$$

In the chain (4.8), we have used the identity $n_i n_i = 1$, the divergence theorem, and the relation $n_{i,i} = (N-1)K$ on Γ_u. From (4.8), we conclude that

(4.9) $$S'(u) = -(N-1) \oint_{\Gamma_u} \frac{K \mathrm{d}s}{|\mathrm{grad} u|}.$$

A second differentiation may be computed using the above device, i.e. we represent again $S'(u)$ as an integral over Ω_u. We have

$$
\begin{aligned}
S'(u) - S'(0) &= (N-1) \oint_{\partial\Omega_u} \frac{K \, ds}{|\mathrm{grad} u|} = \\
&= (N-1) \oint_{\partial\Omega_u} \frac{K n_i n_i}{|\mathrm{grad} u|} ds = \\
&= (N-1) \int_{\Omega_u} \left(\frac{K n_i}{|\mathrm{grad} u|} \right)_{,i} dx = \\
&= (N-1) \int_0^u \oint_{\Gamma_\eta} \left(\frac{n_i K}{|\mathrm{grad} u|} \right)_{,i} |\mathrm{grad} u|^{-1} ds \, d\eta
\end{aligned}
\tag{4.10}
$$

from which we obtain

$$
S''(u) = (N-1) \oint_{\Gamma_u} \left\{ \frac{n_i K}{|\mathrm{grad} u|} \right\}_{,i} |\mathrm{grad} u|^{-1} ds. \tag{4.11}
$$

With

$$
\left\{ \frac{n_i K}{|\mathrm{grad} u|} \right\}_{,i} = \frac{(N-1) K^2}{|\mathrm{grad} u|} + \frac{n_i K_{,i}}{|\mathrm{grad} u|} - \frac{n_i u_{,ki} u_{,k} K}{|\mathrm{grad} u|^3}, \tag{4.12}
$$

the integral in (4.11) is split into three terms. We first evaluate the second term of that integral. With

$$
\begin{aligned}
(N-1) n_i K_{,i} = n_i n_{j,ji} &= \left(n_i \frac{\partial}{\partial x_j} - n_j \frac{\partial}{\partial x_i} \right) n_{j,i} + n_j n_{j,ii} = \\
&= \left(n_i \frac{\partial}{\partial x_j} - n_j \frac{\partial}{\partial x_i} \right) n_{j,i} - n_{i,j} n_{i,j},
\end{aligned}
\tag{4.13}
$$

we have

$$
\begin{aligned}
& (N-1) \oint_{\Gamma_u} \frac{n_i K_{,i}}{|\mathrm{grad} u|^2} ds = \\
&= \oint_{\Gamma_u} |\mathrm{grad} u|^{-2} \left(n_i \frac{\partial}{\partial x_j} - n_j \frac{\partial}{\partial x_i} \right) n_{j,i} ds - \\
&- \oint_{\Gamma_u} \frac{n_{i,j} n_{i,j}}{|\mathrm{grad} u|^2} ds \\
&= - \oint_{\Gamma_u} n_{j,i} n_i \{ |\mathrm{grad} u|^{-2} \}_{,j} ds - \oint_{\Gamma_u} \frac{n_{i,j} n_{i,j}}{|\mathrm{grad} u|^2} ds
\end{aligned}
\tag{4.14}
$$

For the last step in (4.14), we have used the divergence theorem and the identitiy $n_{j,i}n_j = 0$. The various expressions involved in (4.14) are easily computed using the identity (4.3) together with

$$n_i = -\frac{u_{,i}}{\sqrt{u_{,k}u_{,k}}}, \tag{4.15}$$

and

$$n_{i,j} = -\frac{u_{,ij}}{q} + \frac{u_{,kj}u_{,k}u_{,i}}{q^3}. \tag{4.16}$$

We obtain after some reduction

$$\begin{aligned}(N-1)\oint_{\Gamma_u}\frac{n_iK_{,i}}{|\text{grad} u|^2}\mathrm{d}s =\\ = \oint_{\Gamma_u} q^{-8}[q^2u_{,ij}u_{,j}u_{,ik}u_{,k} - (u_{,\ell k}u_{,\ell}u_{,k})^2]\mathrm{d}s -\\ -(N-1)^2\oint_{\Gamma_u}\frac{K^2\mathrm{d}s}{|\text{grad} u|^2} + 2(N-2)\oint_{\Gamma_u}\frac{J}{|\text{grad} u|^2}\mathrm{d}s.\end{aligned} \tag{4.17}$$

From (4.11), (4.12), (4.17), using

$$-\frac{n_iu_{,k}u_{,ik}}{q^4} = (N-1)K\frac{g(q^2)}{q^2G(q^2)} \tag{4.18}$$

on solutions of (1.1), we obtain finally

$$\begin{aligned}S''(u) = \oint_{\Gamma_u} q^{-8}[q^2u_{,ij}u_{,j}u_{,ik}u_{,k} - (u_{,\ell k}u_{,\ell}u_{,k})^2]\mathrm{d}s\\ + 2(N-2)\oint_{\Gamma_u} J|\text{grad} u|^{-2}\mathrm{d}s + (N-1)^2\oint_{\Gamma_u}\frac{K^2g(q^2)}{q^2G(q^2)}\mathrm{d}s.\end{aligned} \tag{4.19}$$

Since the level sets Γ_u are convex, we have $J \geq 0$. Moreover we have

$$\begin{aligned}q^2u_{,ij}u_{,j}u_{,ik}u_{,k} - (u_{,\ell k}u_{,\ell}u_{,k})^2\\ = \frac{1}{2}(u_{,ij}u_{,j}u_{,k} - u_{,ik}u_{,k}u_{,j})(u_{,ij}u_{,j}u_{,k} - u_{,ik}u_{,k}u_{,j}) \geq 0.\end{aligned} \tag{4.20}$$

This leads to the estimate

$$\begin{aligned}S''(u) \geq (N-1)^2\oint_{\Gamma_u}\frac{K^2g(q^2)}{q^2G(q^2)}\mathrm{d}s \geq\\ \geq (N-1)^2\varepsilon\oint_{\Gamma_u}\frac{K^2}{|\text{grad} u|^2}\mathrm{d}s,\end{aligned} \tag{4.21}$$

where $\mathcal{E}$ is defined in (4.2). The proof of Theorem 4 follows from (4.8), (4.9), (4.21), and the Schwarz inequality:

$$\begin{aligned}\frac{S''(u)S(u)}{\mathcal{E}(N-1)^2} &\geq \oint_{\Gamma_u} \frac{K^2}{|\text{grad}u|^2} ds \oint_{\Gamma_u} ds \geq \\ &\geq \left\{\oint_{\Gamma_u} \frac{K}{|\text{grad}u|} ds\right\}^2 = \left(\frac{S'(u)}{N-1}\right)^2,\end{aligned} \tag{4.22}$$

or

$$S''S - \mathcal{E}S'^2 \geq 0. \tag{4.1}$$

In the particular case of the conformal capacity problem (3.1), (3.2), we have $\mathcal{E} = \frac{1}{N-1}$, and we conclude then that

$$[\log S(u)]'' \geq 0 \quad \text{if} \quad N = 2, \tag{4.23}$$

or

$$\left\{[S(u)]^{\frac{N-2}{N-1}}\right\}'' \geq 0 \quad \text{if} \quad N \geq 3. \tag{4.24}$$

The above convexity properties may also be expressed as follows:

$$S(u) \leq S_0^{1-u} S_1^u \quad \text{if} \quad N = 2, \tag{4.25}$$

or

$$[S(u)]^{\frac{N-2}{N-1}} \leq (1-u)S_0^{\frac{N-2}{N-1}} + uS_1^{\frac{N-2}{N-1}} \quad \text{if} \quad N \geq 3. \tag{4.26}$$

The particular case $N = 2$ has already been established by Longinetti in [8]. These upper bounds for $S(u)$ may be complemented by the following lower bounds

$$\log S(u) \geq \begin{cases} \log S_0 + \frac{S'(0)}{S_0} u \\ \\ \log S_1 + \frac{S'(1)}{S_1}(1-u) \end{cases} \quad \text{if } N = 2, \tag{4.27}$$

or

$$[S(u)]^{\frac{N-2}{N-1}} \geq \begin{cases} S_0^{\frac{N-2}{N-1}} \left[1 + \frac{N-2}{N-1} \frac{S'(0)}{S_0} u\right] \\ \\ S_1^{\frac{N-2}{N-1}} \left[1 - \frac{N-2}{N-1} \frac{S'(1)}{S_1}(1-u)\right] \end{cases} \quad \text{if } N \geq 3. \tag{4.28}$$

Combining (4.9) with (4.27b) or (4.28b), we obtain the following lower bound for the maximum value of $|\text{grad}u|$ in the conformal capacity problem (3.1), (3.2):

$$
(4.29) \qquad q_1 \geq \begin{cases} \dfrac{2\pi}{S_1 \log \frac{S_0}{S_1}} & \text{if } N = 2, \\ \dfrac{M_1}{S_1^{1/2}[S_0^{1/2} - S_1^{1/2}]} & \text{if } N = 3. \end{cases}
$$

In (4.29b), $M_1 := \oint_{\Gamma_1} K \mathrm{d}s$ is the Minkowski number of the convex region enclosed by Γ_1.

To conclude this paper, let us mention that similar convexity properties for a variety of functionals defined on the level sets of u may be derived in the same way. For istance we have for the volume $V(u)$ defined in (3.30)

$$
(4.30) \qquad V'(u)V''(u) - \frac{2}{N}[V''(u)]^2 \geq 0
$$

on solutions of the conformal capacity problem (3.1), (3.2), or

$$
(4.31) \qquad [\log V'(u)]'' \geq 0 \quad \text{if} \quad N = 2,
$$

$$
(4.32) \qquad \left\{[V'(u)]^{\frac{N-2}{N}}\right\}'' \geq 0 \quad \text{if} \quad N \geq 3.
$$

The two-dimensional case has already been established by Longinetti in [8].

ACKNOWLEDGMENTS. This paper was completed during the summer 1988, while the first author was visiting the FIM (Forschungsinstitut für Mathematik der ETH-Zürich, Switzerland).

REFERENCES

[1] G. D. ANDERSON: Derivatives of the conformal capacity of extremal rings. *Ann. Acad. Sci. Fenniciae*, Series A. 1, **10** (1985), 29-46.

[2] C. BORELL: Capacitary inequalities of Brunn-Minkoswski type. *Math. Ann.*, **263** (1983), 179-184.

[3] L. A. CAFFARELLI and J. SPRUCK: Convexity properties of solutions to some classical variational problems. *Comm. Part. Diff. Eqs.*, **7** (1982), 1337-1379.

[4] E. HOPF: Elementare Bemerkung über die Lösung partieller Differentialgleichung zweiter Ordnung von elliptischen typus. *Berlin , Sber. Preuss. Akad. Wiss.*, **19** (1927), 147-152.

[5] E. HOPF: A remark on elliptic differential equations of the second order. *Proc. Amer. Math. Soc.*, **3** (1952), 791-793.

[6] B. KAWOHL: Rearrangements and convexity of level sets in PDE. *Lecture Notes in Mathematics*, **1150** (1985).

[7] J. L. LEWIS: Capacitary functions in convex rings. *Arch. Rat. Mech. Anal.*, **66**, (1977), 201-224.

[8] M. LONGINETTI: Some isoperimetric inequalities for the level curves of capacity and Green's functions on convex plane domains. *SIAM J. Math. Anal.*, **19** (1988), 377-389.

[9] L. E. PAYNE: Some applications of «best possible» maximum principles in elliptic boundary value problems. *Pitman, Research Notes in Math.*, **101** (1984), 286-313.

[10] L. E. PAYNE and G. A. PHIIIPPIN: On some maximum principles involving harmonic functions and their derivatives. *SIAM J. Math. Anal.*, **10**, (1979), 96-104.

[11] L. E. PAYNE and G. A. PHIIIPPIN: Some isoperimetric inequalities for capacity, polarization, and virtual mass. *Appl. Anal.*, **23** (1986), 43-61.

[12] M. H. PROTTER and H. F. WEINBERGER: *Maximum principles in differential equations.* Springer Verlag, 1984.

[13] R. SPERB: *Maximum principles and their applications.* Academic Press, Math. in Sci. and Eng., 157, 1981.

SOME GEOMETRIC PROPERTIES OF SOLIDS IN COMBUSTION

G. KEADY - I. STAKGOLD*

0. INTRODUCTION

Under certain assumptions on the governing chemistry, we shall prove two types of result for the combustion of a porous solid by a diffusing gas:

(a) If the solid is initially convex, it remains convex until fully consumed.

(b) Of all solids of equal volume, the ball is slowest to be consumed.

Our results are based on a pseudo-steady-state isothermal model in which the reaction is distributed throughout the solid rather than confined to the surface of a shrinking solid core. Starting from a homogeneous solid occupying the bounded domain Ω, we allow gas to diffuse through the solid and react with it, the gas being supplied continuously through the boundary. The reaction rate is taken to be of first order in the gas concentration C and of quite general form $f(S)$ in the solid concentration S, where $f(S)$ behaves like S^m for small S.

Since the gas concentration is largest at the boundary, the solid will be consumed more quickly near the boundary than in the interior. If $m \geq 1$, the solid concentration remains positive for all time, although, of course, S tends to zero as t tends to infinity. In the more interesting case where $m < 1$ (which occurs in many practical applications such as those involving the popular Sohn-Szekely grain-pellet models), the solid will be completely consumed in finite time. The consumption progresses in two stages: in the first stage, the solid concentration decreases at different rates in Ω until S vanishes on $\partial\Omega$ (at a time we label t_-); in the second stage, the solid occupies a shrinking domain $D(t)$ in which its concentration is positive but varying with position (vanishing on $\partial D(t)$). At a time t_1 the solid is completely consumed (full conversion in finite time).

* Supported by NSF Grant # DMS 8600323.

We shall be interested principally in the case $m < 1$. In the first section we recapitulate some of the results found in Stakgold and McNabb [1984], including an explicit expression for t_1. In the second section we show that if Ω is convex, $D(t)$ is convex for all times $t, t_- < t < t_1$. Thus convexity is preserved in combustion. Our proof depends on earlier work in Kawohl [1985a] and Kennington [1985]. In Section 3 we prove the following isoperimetric result. Among domains of equal volume, the ball converts the solid to products at the slowest rate. In particular, for the case $m < 1$, the time to full conversion is largest for the ball.

An expanded version of this paper with additional details appeared as a technical report (Keady and Stakgold [1987]) of the University of Western Australia with an appendix by Keady and Wynter.

1. THE DIFFERENTIAL EQUATIONS DESCRIBING THE PROBLEM

The pseudo-steady-state combustion problem

A diffusing gas reacts isothermally with a porous solid occupying a domain Ω with smooth boundary. In typical problems the nondimensional porosity ϵ is small and can be neglected. (See, for istance, Stakgold [1982], Stakgold, Bischoff and Gokhale [1983].) Mass balances for the (immobile) solid and the (diffusing) gas concentrations then yield

$$S_t = -f(S)C \qquad x \in \Omega, t > 0, \tag{1.1a}$$

$$-\Delta C = -\varphi^2 f(S)C = \varphi^2 S_t \qquad x \in \Omega, t > 0, \tag{1.1b}$$

with side conditions

$$S(x,0) = 1, \quad x \in \Omega, \tag{1.1c}$$

$$C(\partial\Omega, t) = 1 \quad \text{for} \quad t > 0. \tag{1.1d}$$

Here φ^2 is Thiele's modulus measuring the ratio of reaction to diffusion rates and $f(S)$ is defined for $S \geq 0$ and has the properties

$$(P) \qquad \begin{aligned} &f(0) = 0, \quad f(1) = 1, \quad f \text{ increasing for } \quad S \geq 0, \\ &\lim_{S \to 0+} f(S)S^{-m} = a \quad \text{where } 0 < a < \infty, m > 0. \end{aligned}$$

In many problems $f(S) = S^m$ on $0 \leq S \leq 1$, but we shall not restrict ourselves to that case.

We note first that the initial value $C(x,0)$ is determined by the problem and cannot be imposed. Indeed it is clear from problem (1.1) that $C_0(x) = C(x,0)$ satisfies

$$-\Delta C_0 = -\varphi^2 C_0, \quad x \in \Omega, \quad C_0(\partial\Omega) = 1, \tag{1.2}$$

which has one and only one solution $C_0(x)$. Moreover $0 < C_0(x) < 1$ in Ω and C_0 has a positive normal derivative on the boundary.

In Stakgold [1982] it is shown that problem (1.1) has one and only one solution $(S(x,t), C(x,t))$ which tends to the steady state $(0,1)$ as t tends to infinity, with S descreasing in t and C increasing in t.

If the diffusion rate is much greater than the reaction rate we can set $\varphi^2 = 0$; then $C \equiv 1$ for all x and t, and S becomes independent of x, satisfying the ordinary differential equation

$$S_t = -f(S), \quad t > 0; \quad S(0) = 1. \tag{1.3}$$

The nature of the solution is simple indeed. If $S > 0$, we can divide by $f(S)$ and separate variables to obtain

$$I(S) = \int_S^1 \frac{d\sigma}{f(\sigma)} = t. \tag{1.4}$$

There are two distinct cases according as $I = I(0)$ is infinite or finite. If I is infinite (that is, $m \geq 1$), then equation (1.4) has a unique solution $S(t)$ which is positive for all t and tends to zero as t tends to infinity. If, however, $m < 1$ then I is finite and equation (1.4) has a solution $S(t)$ which is positive for $t < I$. Since $S(I-) = 0$, if follows that the solution $S(t)$ of problem (1.3) vanishes for $t \geq I$. Thus the solid is consumed in the finite time I. It will be convenient to introduce the conversion $\chi = 1 - S$ which, for the case $\varphi^2 = 0$, is given by a definite function $A(t)$ taking on the value unity for $t \geq I$. For $t < I, A$ is defined by

$$I(1 - A(t)) = t,$$

or equivalently

$$A'(t) = f(1 - A(t)), \quad A(0) = 0.$$

We see that

$$A(0) = 0; \quad A'(t) > 0, \quad A''(t) < 0 \quad \text{for} \quad 0 < t < I.$$

Thus, $A(t)$ is concave for $t \geq 0$ and strictly concave for $t < I$.

Let us now return to the case $\varphi^2 > 0$ for which there is spatial variation and $C < 1$ in Ω. From equation (1.1 a) and the initial condition we find immediately that the conversion

$$\chi(x,t) = 1 - S(x,t)$$

can be expressed in terms of $C(x,t)$ by

$$\chi(x,t) = A\left(\int_0^t C(x,\tau)\,\mathrm{d}\tau\right), \tag{1.5}$$

where A is the function defined in the previous paragraph. The argument of A in equation (1.5) is the cumulative concentration introduced by McNabb [1975]. Let us introduce the integrated deviation between the steady-state and transient values of C :

$$\eta(x,t) = \int_0^t (1 - C(x,\tau))\,\mathrm{d}\tau. \tag{1.6}$$

Clearly $0 < \eta < t$ and, in terms of η, equation (1.5) can be written as

$$\chi(x,t) = A(t - \eta(x,t)). \tag{1.7}$$

By integrating equation (1.1 b) from zero to t, we obtain

$$-\Delta\eta = \varphi^2\chi = \varphi^2 A(t-\eta); \quad \eta(\partial\Omega, t) = 0. \tag{1.8}$$

We have therefore reduced the solution of the vector nonlinear problem (1.1) to the scalar problem (1.8) for η, from which C and S are easily recovered. Since $A \le 1$, we have $\eta \le \varphi^2 w(x)$, where $w(x)$ is the solution of the *torsion* problem

$$-\Delta w = 1, \quad x \in \Omega; \quad w(\partial\Omega) = 0. \tag{1.9}$$

It follows from the properties of A and the uniform boundedness of η that $\eta(x,t)$ is increasing in time and that

$$\lim_{t\to\infty} \eta(x,t) = \varphi^2 w(x), \quad \text{uniformly in} \quad \bar{\Omega}.$$

An immediate consequence of equations (1.8), (1.7) and the properties of A is that full conversion in finite time occurs in problem (1.1) if and only if $m < 1$. *We confine ourselves to that case.* Then full conversion occurs precisely at the time t_1, characterized by

$$\min_{x\in\bar{\Omega}}(t_1 - \eta(x,t_1)) = I,$$

or

$$t_1 = I + \max_{x \in \bar{\Omega}} \eta(x, t_1).$$

For $t \geq t_1$, equation (1.8) shows that $\eta = \varphi^2 w(x)$ so that

$$t_1 = I + \varphi^2 \|w\|_\infty, \tag{1.10}$$

where $\|\cdot\|_\infty$ stands for the supremum norm. For $t < t_- = I$, the solid concentration is positive throughout Ω. At $t = I, S(\partial\Omega, I) = 0$ and there is full conversion at the boundary. For t in the range, $I < t < I + \varphi^2 \|w\|_\infty$, we have $S(x,t) > 0$ in a subdomain $D(t)$ of Ω which is then the *domain occupied by the solid at time* t. Of course $S(x,t) < 1$ in $D(t)$ so that our solid has been partly converted in $D(t)$ and fully converted in $\Omega \backslash D(t)$. On $\partial D(t)$ the solid concentration vanishes. Thus, $\partial D(t)$ is the level surface for $\eta(x,t)$ given by

$$\eta(x,t) = t - I \quad (t \text{ fixed}, I < t < I + \varphi^2 \|w\|_\infty).$$

This identification of $\partial D(t)$ as a level surface for the solution of the scalar problem (1.8) is the pivotal step that enables us to use the theorems of Kawohl and Kennington to prove the convexity of $D(t)$.

Preliminary mathematical results concerning problem (1.8)

The results of this subsection apply more generally than merely to problem (1.8). Semilinear elliptic problems of the form

$$-\Delta \eta = h(\eta), \quad \eta(\partial\Omega) = 0, \tag{1.11}$$

arise in numerous applications. Throughout this subsection we suppose that h is bounded on $\mathbf{R}$, nonnegative, and nonincreasing. The maximum principle can be used to establish the following

EXISTENCE AND UNIQUENESS THEOREM. *Suppose h is, in addition to the above, Lipschitz continuous with a uniform bound on its Lipschitz constant. Then there exists a unique solution $\eta \in C^{2,\alpha}(\Omega)$ where α is any number in* $0 \leq \alpha < 1$.

Various approaches to existence are possible: variational, positive operators, etc. Uniqueness follows from the Maximum Principle.

COMPARISON THEOREM. *Suppose* $h_1 \leq h_2$ *everywhere and* $\Omega_1 \subseteq \Omega_2$. *Then the solutions satisfy* $\eta_1 \leq \eta_2$ *on* Ω_1.

The proof uses the Maximum Principle.

We now specialize to

$$h(\eta) = A(t - \eta),$$

and functions A satisfying $A(s) = 0$ for $s \leq 0$ and A nondecreasing and bounded on $(0, \infty)$. New questions arise concerning how the solution $\eta(x, t)$ depends on the nonnegative parameter t. Of course the comparison theorem above implies that η is an increasing function of t.

MAXIMUM BOUND. *For any solution of Problem (1.8),*

$$0 < \eta(x, t) < t \quad \text{in} \quad \Omega \times (0, \infty).$$

PROOF. That $\eta > 0$ is obvious from the fact that η is superharmonic. Let $\psi = t - \eta$. Then

$$\Delta \psi = \varphi^2 A(\psi), \qquad \psi(\partial\Omega) = t.$$

We have, by an intermediate value theorem,

$$-\Delta \psi + \varphi^2 A'(\tilde{\psi}) \psi = 0, \quad \psi(\partial\Omega) > 0,$$

for some $\tilde{\psi} \in (0, \psi)$. Thus $\psi > 0$ in Ω.

We remark that we have not used any concavity hypotheses on A in this subsection.

2. CONCAVITY PROPERTIES AND CONVEXITY OF LEVEL SURFACES

Concavity properties in x at fixed t.

Let Ω be convex. The convexity properties of the solutions η to semilinear Dirichlet problems of the form

$$-\Delta \eta = h(\eta); \quad \eta(\partial\Omega) = 0,$$

have been studied in several recent papers. These include Kennington [1985], Caffarelli and Friedman [1985], Keady [1985], Kawohl [1985 a].

Let η solve problem (1.8). Let g be an increasing function defined on $[0, \eta_{\max})$ with $g(0) = 0$. (In our applications g will also be concave there.) Then $v = g(\eta)$ solves

$$-\Delta v = g'\varphi^2 A(t-\eta) - \frac{g''}{(g')^2}|\nabla v|^2; \quad v(\partial\Omega) = 0,$$

where η has been expressed in terms of v.

Let Ω be convex. That the level curves of $\eta(\cdot, t)$ enclose convex sets can be established by proving, for suitable g, that $g(\eta(\cdot, t))$ is concave over Ω. The main result is the following theorem.

THEOREM 2.1. *Let Ω be a bounded convex domain with C^2 boundary. Let η be a positive solution in $C^2(\Omega) \cap C(\bar{\Omega})$ of Problem (1.8). Then*

(i) $\sqrt{\eta}$ is concave onver Ω.

(ii) $g_2(\eta)$ is concave over Ω where

$$g_2(\eta) = \int_0^{\eta} (\tilde{A}(t,s))^{-1/2} \mathrm{d}\, s, \tilde{A}(t,s) = \int_0^{s} A(t-\sigma) \mathrm{d}\, \sigma.$$

COROLLARY. Under the same assumptions on Ω and η as in Theorem 2.1, the level surfaces of η enclose convex sets and $D(t)$ is convex for each t.

The Corollary is an immediate consequence of Theorem 2.1 (i) or (ii).

We will prove Theorem 2.1 (i) using techniques due to Kennington [1985]. As usual for this sort of question we need two results: an Interior Concavity Lemma and a Boundary Concavity Lemma.

Before presenting these Lemmas we begin by noting that the concavity properties for η given in this section can be obtained by reasonably straighforward adaptations of theorems for elliptic problems already in the literature.

Indeed two special cases of (1.8) are already available in the literature Both special cases are of the type

$$-\Delta \eta = a - b\eta; \quad \eta(\partial\Omega) = 0,$$

with $a > 0$ and $b \geq 0$. Kennington's [1985] results establish that $\sqrt{\eta}$ is concave for this type of problem.

The two special cases of (1.8) that are of this type are as follows.

Firstly,

$$\eta(\cdot, t) \sim (1 - C_0(\cdot))t \quad \text{as} \quad t \to 0+,$$

and C_0 solves problem (1.2). Let $\eta_0 = 1 - C_0$. Then

$$-\Delta \eta_0 = \varphi^2(1 - \eta_0); \quad \eta_0(\partial\Omega) = 0,$$

which is the case $a = b = \varphi^2$ in the above. Thus $\sqrt{\eta_0}$ is concave.

Also, for $t > t_1, \eta(\cdot, t) = \varphi^2 w(\cdot)$ where w is the torsion function, the case $a = \varphi^2, b = 0$ in the above. Hence $\sqrt{w}$ is concave. This result was first proved, for two spaces dimensions, by Makar-Limanov [1971], and, for general dimensions, by Kennington [1985].

We now prepare for the proof of Theorem 2.1.

INTERIOR CONCAVITY LEMMA. *Let v satisfy*

$$-\Delta v = b(v, \nabla v) \ in\, \Omega,$$
$$v = 0 \ in\, \partial\Omega.$$

Suppose that it is known that any positive solution is such that

$$(v, \nabla v) \in \mathcal{V},$$

where $\mathcal{V}$ is a convex subset of $\mathbb{R}^{n+1}$. Suppose that, on $\mathcal{V}, b$ is non-negative, and that, at each fixed q such that some $(v, q) \in \mathcal{V}$,

(i) $b(\cdot, q)$ is strictly decreasing in its first argument (v), and

(ii) $\frac{1}{b(\cdot,q)}$ is convex in its first argument.

Suppose, in addition, that for all $z_0 \in \partial\Omega, z \in \Omega$ and $\tau \in [0, 1]$,

$$v((1 - \tau)z_0 + \tau z) \geq \tau v(z). \tag{2.1}$$

Then v is concave over Ω.

PROOF. This is a special case of Theorem 3.1 of Kennington [1985].

In order to prove Theorem 2.1 (i), it suffices to verify the hypotheses in the Lemma for $v = \eta^2$. As a first step we prove that (2.1) holds.

BOUNDARY CONCAVITY LEMMA. *Here we do not need to assume that Ω is convex. Let $\Omega \subset \mathbb{R}^n$ with the origin $\in \partial\Omega$. Suppose that Ω is star-shaped with respect to the origin. Suppose also that the interior sphere condition is satisfied at every point of the boundary of Ω.*

Then, for any solution of Problem (1.8)

$$\sqrt{\eta(\tau z)} > \tau\sqrt{\eta(z)} \quad \text{for all} \quad z \in \Omega, \tau \in (0, 1).$$

PROOF. A short calculation gives

$$(-\Delta + \varphi^2 A'(t-\eta))\,(z \cdot \nabla\eta - 2\eta) \leq 0 .$$

By hypothesis Ω is star-shaped with respect to the origin so that $(z \cdot \nabla\eta - 2\eta) < 0$ on $\partial\Omega$ and hence, by the Maximum Principle, throughout Ω.

Let $z \in \Omega$ and write $\eta(\tau) = \eta(\tau z)$. The last sentence of the preceding paragraph yields

$$\frac{\mathrm{d}}{\mathrm{d}\tau}\,\frac{\eta^{1/2}(\tau)}{\tau} < 0 .$$

Integrating from τ to 1 gives

$$\eta^{1/2}(1) - \frac{\eta^{1/2}(\tau)}{\tau} < 0 \quad \text{for} \quad 0 < \tau < 1 .$$

This is precisely the result claimed in the Lemma.

PROOF OF THEOREM 2.1 (i). If we set $\eta = v^2$ the problems for η and for v are related by

$$b(v,q) = \frac{2q^2 + \varphi^2 A(t - v^2)}{2v} ,$$

where q^2 is the square of the Euclidean length of vector q. Hypotheses (i) and (ii) on b are satisfied:

For hypothesis (i) on b we note that $\frac{1}{v}$ is strictly decreasing and $2q^2 + \varphi^2 A(t - v^2)$ is non-increasing in v.

For hypothesis (ii) on b we note that

$$\frac{\mathrm{d}^2}{\mathrm{d}v^2}\left(\frac{v}{2q^2 + \varphi^2 A(t-v^2)}\right) = \\ = \frac{\varphi^2((2q^2 + \varphi^2 A)\,(6vA' - 4v^3 A'') + 4v^3 \varphi^2 A'^2}{(2q^2 + \varphi^2 A)^3} .$$

The properties of A then ensure the hypotheses on b are satisfied. We have therefore verified all the hypotheses of the Interior Concavity Lemma and (2.1). Therefore v is concave over Ω and Theorem 2.1 (i) has been proved.

PROOF OF THEOREM 2.1 (ii). The observation that the function g_2 was appropriate to consider in problems like ours is due to Caffarelli and Friedman [1985]. The result here, in n dimensions, follows from Kawohl [1985 a, Corollary 2 of Theorem 1].

REMARK. The proof that the result of part (ii) of the theorem is stronger than the result of part (i) is easy. We now present it. The starting point is the following elementary lemma:

LEMMA. *Lemma v be a concave function of x in Ω. Let g be an increasing concave function on $[0, v_{\max}]$. Then $g \circ v$ is concave over Ω.*

Define g on $[0, g_2(\eta_{\max})]$ by

$$\sqrt{\eta} = g(g_2(\eta)).$$

To show that $g_2(\eta)$ concave implies $\sqrt{\eta}$ concave all that is needed is to show that g is increasing and concave. Write the definition of g as

$$s = g(g_2(s^2)).$$

Then

$$1 = \frac{g'(g_2(s^2)) \cdot 2s}{\sqrt{\tilde{A}(t, s^2)}},$$

so g is increasing. Next

$$(g'(g_2(s^2))^2 = \frac{\tilde{A}(t, s^2)}{4s^2},$$

so

$$\frac{4sg'g''}{\sqrt{\tilde{A}(t, s^2)}} = \frac{A(t - s^2)}{2s} - \frac{\tilde{A}(t, s^2)}{2s^3}.$$

Since $A(s)$ is an increasing function of s,

$$\tilde{A}(t, \eta^2) = \int_0^{\eta} A(t - \hat{\eta}) \, d\hat{\eta} \geq \eta A(t - \eta),$$

so that $g'' \leq 0$ as required.

Joint concavity properties in (x, t)

Even without any convexity restriction on Ω we have the following:

THEOREM 2.2. *For $0 < t < t_1$*

(i) η is increasing in t at fixed x. Also $0 < \eta_t < 1$.

(ii) η is concave in t at fixed x.

PROOF. This proof is straighforward and can be found in Keady and Stakgold [1987].

For the one-dimensional domain $\Omega = (-1, 1)$ we have

THEOREM 2.3. *At each fixed value of* $t \geq 0, \eta(\cdot, t)$ *is concave in* x *and the function* η *is quasiconcave jointly in* (x, t).

PROOF. The first part is obvious from the problem (1.8) and the coincidence of superharmonic and concave functions in one space dimension. For the second part define

$$T = -\eta_x^2 \eta_{tt} + 2\eta_x \eta_t \eta_{xt} + \eta_t^2 A(t - \eta).$$

We must show that $T \geq 0$ everywhere. Clearly $T = 0$ for $t \geq t_1$ so we need only consider $0 < t < t_1$. Using Theorem 2.2 it is seen that the first and the last terms in the expression for T are both positive.

Thus the result is easily established if it is shown that η_x and η_{xt} have the same sign. From problem (1.8), since η is positive, symmetric and concave, η_x is negative in $0 < x < 1$. Similarly, using both parts of the result in Theorem 2.2 (i), η_t is also positive, symmetric and concave and so η_{xt} is negative in $0 < x < 1, 0 < t < t_1$.

The question arises as to whether some power of η is jointly concave in (x, t). Only partial answers are available. The case $m = 0$, that is, $f(S) = 1$, has been investigated in one dimension by Keady and Wynter in the Appendix to Keady and Stakgold [1987]. Some additional results are also given for the two-dimensional case with $m = 1$.

3. ISOPERIMETRIC RESULTS

In this section we shall compare the combustion properties of a solid of arbitrary shape Ω with that of a ball of the same volume. We use the familiar language of three dimensions although the mathematical results apply to any number of dimensions.

Our first result follows immediately from the explicit formula (1.9) which gives the time of full conversion (where $m < 1$) as

$$t_1 = I + \varphi^2 ||w||_\infty,$$

where $||w||_\infty$ is the maximum of the solution of the torsion problem for the domain Ω. It is known (Bandle [1980], Payne [1967]) that, among domains of equal volume, the largest value of $||w||_\infty$ occurs for the ball. This gives the following:

THEOREM 3.1. *Of all domains of equal volume, the time to full conversion is greatest for the ball.*

Next we consider partial conversion. Again, we would like to show that partial conversion is slowest for a ball, a result which will hold for any $m \geq 0$. For a domain Ω, the overall fraction of solid converted at time t is given by

$$\gamma(t) = \frac{1}{|\Omega|} \int_{\Omega} \chi(x,t) \mathrm{d}\, x = \frac{1}{|\Omega|} \int_{\Omega} A(t - \eta(x,t)) \mathrm{d}\, x, \tag{3.1}$$

where η solves problem (1.8).

Denoting by a superscript star, e.g. $\gamma^{\star}$, the similar quantities for the problem (1.8) on a ball of the same volume as Ω, we have the following theorem:

$$\gamma^{\star}(t) \leq \gamma(t), \quad \text{for every} \quad t > 0. \tag{3.2}$$

(This is stated more formally as Theorem 3.2 below.)

The proof is based on using rearrangements. (See, for instance, Bandle [1980], Bandle, Sperb and Stakgold [1984], Kawohl [1985 b], Talenti [1976].) Before proceeding with the proof, we shall collect some known information about rearrangements. Let $\eta(x)$ be the solution of problem (1.8) for some fixed t. Our notation suppresses the dependence of η on t. The distribution function of η is the volume $\mu(a)$ of the region where $t - \eta(x) \leq a$,

$$\mu(a) = \text{meas.}\, \{x \in \Omega \,|\, t - \eta(x) \leq a\}. \tag{3.3}$$

Clearly $\mu(a)$ is defined for $\alpha \leq a \leq t$, where α is the minimum of $t - \eta$ on $\bar{\Omega}$, thus $\mu(a) = 0, \mu(t) = |\Omega|$. Since μ is increasing, its inverse $a(\mu)$ is defined and increasing, $0 \leq \mu \leq |\Omega|$, with $a(0) = \alpha, a(|\Omega|) = t$. The function $a(\mu)$ is the *increasing rearrangement* of $t - \eta(x)$.

A formal calculation (which can be justified in the present context) shows that

$$[a'(\mu)]^{-1} = \int_{\eta = t - a(\mu)} \frac{\mathrm{d}\, s}{|\nabla \eta|}, \quad \mu > 0, \tag{3.4}$$

where the integration is over the level surface $\eta = t - a(\mu)$, that is, the level surface enclosing the volume μ. Its area will be denoted by $s(\mu)$.

On setting

$$E(\mu) = \varphi^2 \int_{\eta \geq t - a(\mu)} A(t - \eta(x)) \mathrm{d}\, x,$$

we find from problem (1.8) that

$$E(\mu) = \int_{\eta = t - a(\mu)} |\nabla \eta| \mathrm{d}\, s.$$

Combining this last equation with equation (3.4) and using first Schwarz's inequality, and then the classical isoperimetric inequality, we obtain

$$E(\mu) \geq s^2(\mu)a'(\mu) \geq \sigma^2(\mu)a'(\mu), \tag{3.5}$$

where $\sigma(\mu)$ is the surface area of a ball of volume μ.

Since

$$E'(\mu) = \phi^2 A(a(\mu)),$$

it follows that

$$E(\mu) = \varphi^2 \int_0^\mu A(a(\nu))\mathrm{d}\nu. \tag{3.6}$$

The unique solution $\eta^\star(x)$ of problem (1.8) for the ball $\Omega^\star$ with $|\Omega^\star| = |\Omega|$ is radially symmetric and inequality (3.5) then becomes an equality. Combining our results, we find

$$\sigma^2(a' - (a^\star)') \leq E - E^\star = \varphi^2 \int_0^\mu (A(a(\nu)) - A(a^\star(\nu)))\mathrm{d}\nu. \tag{3.7}$$

We shall use inequality (3.7) to prove

THEOREM 3.2. *Let $\eta(x)$ be the solution of Problem (1.8) in Ω and $\eta^\star(x)$ be the solution for a ball $\Omega^\star$ of the same volume as Ω. Then the respective conversions (see definition (3.1)) satisfy*

$$\gamma^\star(t) \leq \gamma(t), \quad \text{for every} \quad t > 0.$$

PROOF. It obviously suffices to show that $E(|\Omega|) \geq E^\star(|\Omega|)$. Suppose instead $E(|\Omega|) < E^\star(|\Omega|)$. In view of the equality $E(0) = E^\star(0) = 0$, there must exist a maximal interval $h < \mu \leq |\Omega|$ in which $E(\mu) < E^\star(\mu)$ and $E(h) = E^\star(h)$. Inequality (3.7) then shows that

$$a'(\mu) - a^{\star\prime}(\mu) < 0 \quad \text{on} \quad (h, |\Omega|],$$

so that integration from μ to $|\Omega|$ yields

$$a^\star(\mu) < a(\mu) \quad \text{on} \quad (h, |\Omega|]. \tag{3.8}$$

Since $E(h) = E^\star(h)$, equation (3.6) and the equivalent expression for $E^\star$ give

$$E(\mu) - E^\star(\mu) = \varphi^2 \int_h^\mu (A(a(\nu)) - A(a^\star(\nu)))\,d\nu.$$

The monotonicity of A and the inequality (3.8) then imply that

$$E(\mu) - E^\star(\mu) \geq 0 \quad \text{on} \quad (h, |\Omega|].$$

This contradiction establishes the theorem.

REMARKS

(i) We can also show that $E(\mu) - E^\star(\mu) \geq 0$ for all μ.

(ii) If $\alpha, \alpha^\star$ are the maximal values of η and $\eta^\star$, respectively, then $\alpha^\star \geq \alpha$.

(iii) That A is concave has not been used in Section 3.

(iv) If $m < 1$, the solid is totally consumed in finite time and will occupy a domain $D(t)$ at time t. Let $D^\star(t)$ be the similar domain for a solid initially in the shape of a ball. We conjecture that

$$|D(t)| \leq |D^\star(t)|.$$

Various other isoperimetric inequlities of lesser intuitive appeal are also proved in Keady and Stakgold [1987].

REFERENCES

[Ba80] C. BANDLE: Isoperimetric Inequalities and Applications. *Pitman*, 1980.

[Ba84] C. BANDLE, R. SPERB and I. STAKGOLD: Diffusion-reaction with monotone kinetics. *Nonlinear Analysis*, **8** (1984), 275-293.

[Caf85] L. A. CAFFARELLI and A. FRIEDMAN; Convexity of solutions of semilinear elliptic equations. *Duke Math. J.*, **52** (1985), 431-456.

[Kaw85a] B. KAWOHL: When are solutions to nonlinear elliptic boundary problems convex? *Comm. in P.D.E.*, **10** (1985), 1213-1225.

[Kaw85b] B. KAWOHL: Rearrangements and Convexity of Level Sets in P.D.E. Springer Lecture Notes in Maths, 1150, 1985.

[Kea85] G. KEADY: The power concavity of solutions of some semilinear elliptic boundary value problems. *Bull. Aust. Math. Soc.*, **31**, (1985), 181-184.

[Kea87] G. KEADY and I. STAKGOLD: *Combustion of convex solids*. University of Western Australia, May 1987, with Appendix by G. Keady and P. Wynter.

[Ken85] A. U. KENNINGTON: Power concavity and boundary value problems. *Indiana J. Math. Anal.*, **34** (1985), 687-704.

[Mak71] L. G. MAKAR-LIMANOV: Solution of Dirichlet's problem for the torsion equation in a convex region. *Math. Notes Acad. Sci. U.S.S.R.*, **9** (1971), 52-53.

[Mc75] A. McNabb: Asymptotic behaviour of solutions of diffusion equations. *J. Math. Anal. Appl.*, **51** (1975).

[Pay67] L. E. Payne: Isoperimetric inequalities and their applications. *S.I.A.M. Review*, **9** (1987), 453-488.

[St82] I. Stakgold: Gas-solid reactions. In *Dynamical Systems II*, A. R. Bednarek and L. Cesari eds., 403-417, Academic Press, 1982.

[St83] I. Stakgold, K. B. Bischoff and V. Gokhale: Validity of the pseudo-steady-state approximation. *Int. J. of Eng. Sci.*, **21** (1983), 537-542.

[St84] I. Stakgold and A. McNabb: Conversion estimates for gas-solid reactions. *Mathematical Modelling*, **5** (1984), 325-330.

[St85] I. Stakgold: Reaction diffusion problems in chemical engineering. In *Proceedings of the Conference on Free and Moving Boundary Problems*, Montecatini, 1985.

[Ta76] G. Talenti: Elliptic equations and rearrangements. *Annali Scu. norm. sup.*, Pisa, Ser. 4, **3** (1976), 697.

RADIAL AND NON-RADIAL SOLUTIONS FOR SEMILINEAR ELLIPTIC EQUATIONS ON CIRCULAR DOMAINS

TAKASHI SUZUKI

PART I. GNN THEORY AND RELATED PROBLEMS

1. Introduction

Our aim is to describe some recent study on radial and non-radial solutions for semilinear elliptic equations on circular domains. Thus, let $\Omega \subset \mathbf{R}^n$ be a bounded domain with a smooth boundary $\partial\Omega$. For a C^1-functions $f : \mathbf{R} \to \mathbf{R}$ we consider the semilinear elliptic boundary value problem

$$-\Delta u = f(u), \quad u > 0 \quad (\text{in } \Omega) \quad \text{and} \quad u = 0 \quad (\text{on } \partial\Omega). \tag{1.1}$$

When Ω is a disc: $\Omega \equiv D \equiv \{|x| < 1\} \subset \mathbf{R}^n$, then every solution $u = u(x)$ of (1.1) must be radial i.e., $u = u(|x|)$ and $u_r < 0$ holds for $0 < r = |x| < \mathbf{R}$. This remarkable fact due to [GNN] reduces the problems (1.1) to the one-dimensional problem

$$\begin{aligned} &-\left(\frac{d^2}{dr^2} + \frac{n-1}{r}\frac{d}{dr}\right) u = f(u) \quad (0 < r < R) \quad \text{with} \\ &u'(0) = u(R) = 0. \end{aligned} \tag{1.2}$$

For instance, given a constant $\lambda > 0$ every solution $u = u(x)$ of

$$-\Delta u = \lambda e^u \quad (\text{in } \Omega), \quad u = 0 \quad (\text{on } \partial\Omega) \tag{1.3}$$

is positive, so that the problem is reduced to

$$\begin{aligned} &-\left(\frac{d^2}{dr^2} + \frac{n-1}{r}\frac{d}{dr}\right) u = \lambda e^u \quad (0 < r < R) \quad \text{with} \\ &u'(0) = u(R) = 0, \end{aligned} \tag{1.4}$$

when $\Omega = D$. Thus the following diagram due to [G] and [JL] for (1.4) gives a complete structure of the solution set for (1.3) in this case. Here, we suppose $R = 1$ without loss of generality.

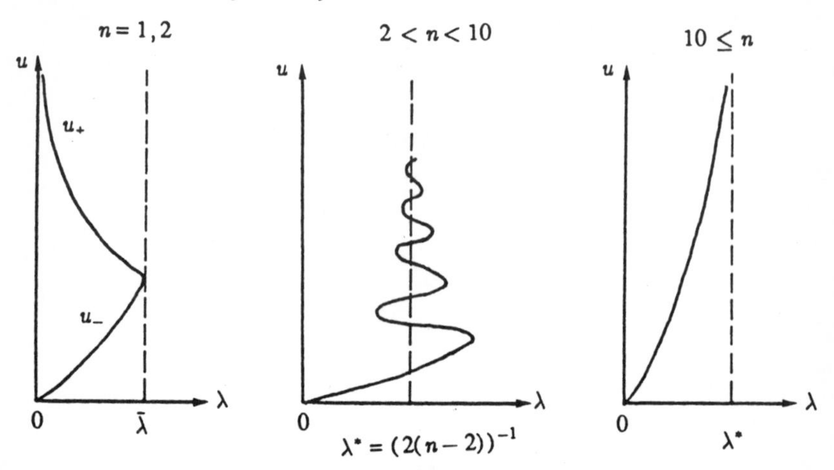

The proof follows the idea of [A] called moving plane method. Namely, we move a hyperplane T along its normal direction. Then, some of its translation T_λ cuts a portion Σ of Ω with the property $\Sigma' \subset \Omega, \Sigma'$ being the reflection of Σ with respect to T_λ. Taking now the reflection u^λ of u with respect to T, we make comparison in Σ' between u and u^λ. Here, important is that the relation $-\Delta u = f(u)$ is invariant under the transformation $u \mapsto u^\lambda$. We note that the Kelvin trasformation has a similar property. The other kind of such a tranformation is utilized in [CNS].

A consequence of [GNN] derived in this way is described as follows: Let $\gamma \in \mathbf{R}^n$ be a unit vector and $T_\lambda = \{x \in \mathbf{R}^n | x \cdot \gamma = \lambda\}$ be a hyperplane, where λ is a real number. We put $\lambda^* = \sup\{\lambda | T_\lambda \cap \Omega \neq \emptyset\}, \lambda_* = \text{Inf}\,\{\lambda | T_\lambda \cap \Omega \neq \emptyset\}$ quad $T = T_{(\lambda_* + \lambda^*)/2}$.

DEFINITION 1. We say that a domain $\Omega \subset \mathbf{R}^N$ is GNN-symmetric with respect to T if the following are satisfied:

1) Ω is symmetric with respect to T.

2) For $\lambda \in ((\lambda_* + \lambda^*)/2, \lambda^*)$, the reflection set G'_λ of $G_\lambda = \{x \in \Omega | x \cdot \gamma > \lambda\}$ with respect to T_λ lies in Ω.

3) For $\lambda \in ((\lambda_* + \lambda^*)/2, \lambda^*), T_\lambda$ is not perpendicular to $\partial\Omega$.

Then , we have

THEOREM 1. ([GNN]). If $\Omega \subset \mathbf{R}^n$ is GNN-symmetric with respect to T, every solution $u = u(x) \in C^2(\Omega) \cap C^0(\bar{\Omega})$ of (1.1) is symmetric with respect to T

and satisfies $\gamma \cdot \nabla u(x) < 0$ for $x \in \Omega$ with $\frac{1}{2}(\lambda_* + \lambda^*) < \gamma \cdot x < \lambda^*$.

In other words, we can examine from this theory when the solution u inherits the symmetry of the domain Ω and how it decreases within Ω. That is, u is symmetric if Ω is GNN symmetric, and decreases along each straight line starting from T in this case. From Theorem 0; the radial property of $u = u(|x|)$ for (1.1) on D and the fact $u_r < 0 (0 < r < R)$ follow immediately.

A natural question is that what happens when GNN-symmetry of Ω is violated and some interesting phenomena are expected. Actually, in [CNS] we have given some other decreasing streamlines of solutions for some symmetric domains having necks. On the other hand, in the annular domain $\Omega = A = \{a < |x| < 1\} \subset \mathbf{R}^n$ non-radial solutions can arise for (1.1).

2. Generation of non-radial solutions

In 1983, [BN] considered the boundary value problem

$$-\Delta u = u^p, \quad u > 0 \quad (\text{in } \Omega), \quad u = 0 \quad (\text{on } \partial\Omega) \tag{2.1}$$

on the annulus $\Omega = \{R_1 < |x| < R_2\} \subset \mathbf{R}^n$. From the variational approach, they observed that non-radial solutions for (2.1) arise as $p \nearrow n^*$, where n^* is the critical Sobolev exponent : $n^* = +\infty$ if $n = 2$ and $n^* = (n+2)/(n-2)$ if $n > 2$.

Their argument is as follows. put $V = H_0^1(\Omega), V_\infty = \{v \in H_0^1(\Omega) | v = v(|x|)\}$,

$$j = \operatorname*{Inf}_{v \in V \setminus \{0\}} ||\nabla V||_{L^2} / ||v||_{L^{p+1}} \quad \text{and} \quad j_\infty = \operatorname*{Inf}_{v \in V_\infty \setminus \{0\}} ||\nabla v|| / ||v||_{L^{p+1}}.$$

There exists a minimizer $u \in V \setminus \{0\}$ of j satisfying (2.1). In fact, we can take a non-negative u and then the Lagrangean multiplier is reduced to 1 after the stretching transformation because of the homogenity property of the nonlinearl-ity $f(t) = t^p$. On the other hand, a minimizer, $u_\infty \in V_\infty \setminus \{0\}$ of j_∞ also solves (2.1), which can be verified through the argument of [KW]. In fact, it holds that

$$\int_{R_1}^{R_2} r^{n-1} \{u'_\infty v' - u_\infty^p v\} d\,r = 0$$

for each $v \in X_\infty$, from which $-\frac{1}{r^{n-1}} \frac{d}{dr} (r^{n-1} \frac{d}{dr} u_\infty) = u_\infty^p$ follows.

Now, what they have shown is that $j < j_\infty$ holds if $p < n^*$ is close to n^*. Then, $u \notin X_\infty$ follows.

Motivated by this, [C] has shown the generation of essentially infinetely many solutions for

$$(2.2) \qquad -\Delta u + u = u^{2\ell+1}, \quad u > 0 \quad (\text{in } \Omega), \quad u = 0 \quad (\text{on} \, \partial\Omega)$$

in $\Omega = \{a < |x| < a + b\} \subset \mathbf{R}^2$ as $a \to +\infty$ with $b > 0$ fixed, where $\ell = 1, 2, \ldots$ Whithout loss of essense, we shall decribe his method for the boundary value problem (2.1) in the same domain $\Omega = \{a < |x| < a + 1\} \subset \mathbf{R}^2$.

Thus, for $k = 1, 2, \ldots,$ we define the operation T_k on $V = H_0^1(\Omega)$, which is nothing but the rotation in θ-direction by $\frac{2\pi}{k}$ of the independent variables $x = re^{i\theta}(r = |x|)$, i. e.,

$$(T_k v)(re^{i\theta}) = v\left(re^{i(\theta + \frac{2}{k})}\right) \quad (a < r < a + b, \theta \in (0, 2\pi)).$$

We put $V_k = \{v \in V | T_k v = v\}$ for $k = 1, 2, \ldots, V_\infty = \{v \in V | v = v(|x|)\}$ and

$$(2.3) \qquad j_k = \inf_{v \in V_k \setminus \{0\}} ||\nabla v||_{L^2} / ||v||_{L^{p+1}} \quad (k = 1, 2, \ldots, \infty).$$

Hence $j_1 = j$ in the previous notation. We first assert that

PROPOSITION 2.1. For $k = 1, 2, \ldots, \infty$, there exists a minimizer $u = v_k \in V_k$ of j_k satisfying (2.1).

This fact, which is never trivial and is proven in [C] originally through the argument on [N], extends to other problems without homogeneous property in their nonlinearities ([SN3]). We shall describe the argument later in Part II and admit the proposition for the moment.

Thus, the following proposition assures us of the generation of non-radial solutions:

PROPOSITION 2.2. Under the operation

$$(\#) \qquad a \to +\infty \quad \text{with} \quad b > 0 \quad \text{fixed},$$

we have $j_\infty \to +\infty$, while j_k is bounded for finite k's.

PROOF. We first show what $j_k \in 0(1)$ if k is finite. To this end, let $\Omega_k = \{re^{i\theta} | a < r < a + b, 0 < \theta < \frac{2\pi}{k}\}$ be the fundamental region. It is easy to see that under the operation $(\#)$, there exist a ball B with radius $C > 0$ independent of a such that its suitable translation B' lies in Ω_k. Taking a non-zero function

$\varphi \in C_0^\infty(\mathbf{R}^2)$ with supp $\varphi \subset B$, we can give its translation φ_1 such that supp $\varphi_1 \subset B'$. Then, let $\varphi_2, \ldots, \varphi_k$ be its rotations in θ-direction by $\frac{2\pi}{k}$. We have $\psi \equiv \varphi_1 + \ldots + \varphi_k \in V_k$, so that

$$j_k \leq J(\psi) \equiv ||\nabla \phi||_{L^2(\Omega)} / ||\phi||_{L^{(p+1)}(\Omega)} = \\ k^{1/2-1/p+1} ||\nabla \phi||_{L^2(B)} / ||\phi||_{L^{p+1}(B)}$$

Hence $j_k \in 0(1)$ follows.

We next show that $j_\infty \to +\infty$ under (#). Thus, we take an arbitrary element $v = v(r) \in V_\infty$. Since $v(r) = -\int_r^{a+b} v'(\rho) \mathrm{d}\rho$, we have

$$v(r^2) \leq \int_r^{a+b} v(\rho)^2 \rho \mathrm{d}\rho \int_r^{a+b} \rho^{-1} \mathrm{d}\rho \leq \\ \leq \int_a^{a+b} v'(\rho)^2 \rho \mathrm{d}\rho r^{-1} \int_a^{a+b} \mathrm{d}\rho \approx ||v||_{L^2}^2 \frac{b}{r},$$

so that

$$\int_a^{a+b} v(r)^{p+1} r \mathrm{d}r \lesssim ||\nabla v||_{L^2}^{p+1} \int_a^{a+b} r^{-\frac{p-1}{2}} \mathrm{d}r \approx ||\nabla v||_{L^2}^{p+1} a^{-\frac{p-1}{2}}.$$

Hence $j_\infty \gtrsim a^{(p-1)/2(p+1)} \to +\infty$ by $p > 1$.

To get more detailed information about non-radial solutions, we have to examine the separation of critical values $\{j_k\}$ for finite k's. The original result by [C] was rather complicated and [K] has refined it through a simpler argument using Steiner's symmetrization. Here, we shall present third argument, which might be simpler even than [K]'s one.

PROPOSITION 2.3. We have

$$j_1 \leq j_2 \leq \cdots \leq j_\infty. \tag{2.4}$$

Furthermore, $j_k < j_\infty$ implies that $j_1 < \ldots < j_k$.

PROOF. For each $v = v(re^{i\theta}) \in V_k$, we set $\tilde{v}(r, \theta') = v(re^{i\theta})$ with $\theta' = k\theta$. Then, the mapping $S_k : v \in V_k \to \tilde{v} \in V_1 = V$ is onto. Further we have

$$||\nabla v||_{L^2}^2 = \int_0^{2\pi} \mathrm{d}\theta \int_a^{a+b} \mathrm{d}r \left\{ r \left(\frac{\mathrm{d}v}{\partial \theta} \right)^2 + r^{-1} \left(\frac{\mathrm{d}v}{\partial r} \right)^2 \right\} \\ = k \int_0^{\frac{2\pi}{k}} \mathrm{d}\theta \int_a^{a+b} \mathrm{d}r \left\{ r \left(\frac{\partial v}{\partial r} \right)^2 + r^{-1} \left(\frac{\partial v}{\partial \theta} \right)^2 \right\} \\ = \int_0^{2\pi} \mathrm{d}\theta' \int_a^{a+b} \mathrm{d}r \left\{ r \left(\frac{\partial \tilde{v}}{\partial r} \right)^2 + k^2 r^{-1} \left(\frac{\partial \tilde{v}}{\partial \theta'} \right)^2 \right\}$$

and

$$\|v\|_{L^{p+1}}^{p+1} = k \int_0^{\frac{2\pi}{k}} d\theta \int_a^{a+b} dr |v|^{p+1} r =$$
$$= \int_0^{2\pi} d\theta' \int_0^{a+b} dr |\tilde{v}|^{p+1} r$$
$$= \|\tilde{v}\|_{L^{p+1}}^{p+1}.$$

Hence for $k = 1, 2, \ldots$

$$j_k = \operatorname*{Inf}_{v \in V \setminus \{0\}} J_k(v). \tag{2.5}$$

where $J_k(v) = \left\{ \int_0^{2\pi} d\theta \int_a^{a+b} dr \left(r \left(\frac{\partial v}{\partial r}\right)^2 + k^2 r^{-1} \left(\frac{\partial v}{\partial \theta}\right)^2 \right) \right\}^{1/2} / \|v\|_{L^{p+1}}$. Since $J_k(v) \leq J_{k+1}(v)$ for each $v \in V$ and $k = 1, 2, \ldots,$ we obtain (2.4).

There exists a minimizer $v_k \in V \setminus \{0\}$ of j_k in (2.5) such that $\|v_k\|_{L^{p+1}} = 1$. For this v_k we have

$$j_{k-1} \leq J_{k-1}(v_k) \leq J_k(v_k) = j_k.$$

Therefore, the relation $j_{k-1} = j_k$ implies $J_k(v_k) = J_{k-1}(v_k)$, and hence $\frac{\partial v_k}{\partial \theta} \equiv 0$. In other words $v_k \in V_\infty$, so that $j_k = j_\infty$ because $u_k = S_k^{-1} v_k \in V_\infty$ is a minimizer of j_k in (2.3). Thus $j_1 < \ldots < j_k$ follows from $j_k < j_\infty$ by an induction.

Under these proposition, we can give the following theorem, where

DEFINITION 2. For each $v \in V, k = \sup\{\ell | v \in V_\ell\}$ is said to be the mode of v. Hence v is radial if its mode is ∞.

THEOREM 2. There always exists a radial, that is of mode ∞, solution for (2.1) on $\Omega = \{a < |x| < a + b\} \subset \mathbf{R}^2$. Furthermore, any k-mode solutions arise under the operation

$$a \to +\infty \quad \text{with} \quad b > 0 \quad \text{fixed}. \tag{\#}$$

PROOF. The first part follows from Proposition 2.1. For each finite k, the relation $j_k < j_\infty$ arises under (#) by Proposition 2.2. We shall show that then the minimizer u_k in Proposition 2.1 is of mode k. In fact, if $u_k \in V_{k'}$, for some

$k' > k$, then we get $j_{k'} = j_k$ in (2.4) and hence $j_{k'} = j_\infty$ by Proposition 2.3. Therefore $j_k = j_\infty$, which contradicts to $j_k < j_\infty$.

It seems to be a quite interesting question whether any k-mode solution of (2.1) is a minimizer of $j_k = \inf_{v \in V_k \setminus \{0\}} \|\nabla v\|_{L^2} / \|v\|_{L^{p+1}}$ or not. If it is true, from the above argument we can deduce that when a k-mode solution for (2.1) arises, then there exists any ℓ-mode solution for $\ell = 1, 2, \ldots, k-1$. On the other hand, the generation of non-radial solutions can be approached from bifurcation theory. See also Part II of the present article.

PART II. RADIAL AND NON-RADIAL SOLUTIONS FOR $\Delta u + \lambda e^u = 0$.

1. Introduction

The nonlinear eigenvalue problem

$$(1.1) \qquad -\Delta u = \lambda e^u \quad (\text{in } \Omega), \quad u = 0 \quad (\text{on } \partial\Omega)$$

arises in a wide variety of situations, e.g., in the theory of thermal self-ignition of a chemically active mixture of gasses [G], in the theory of nonlinear heat generation [KC] and in the study of Riemann surfaces with constant Gaussian curvature [KW], where $\Omega \subset \mathbf{R}^n$ is a bounded domain with a smooth boundary $\partial\Omega$ and λ is a positive constant.

Up to now, many studies have been done about the existence of solutions. We add [MMP], [B2], [SN2, 4] and [S] to the references of [NS] and [SN1]. In the present part we only study (1.1) for circular domains in $\mathbf{R}^2$, that is, $\Omega \equiv \{|x| < 1\} \subset \mathbf{R}^2$ or $\Omega = A \equiv \{a < |x| < 1\} \subset \mathbf{R}^2$, where $0 < a < 1$, and in this section we describe the solution set for $\Omega = D$. The result has been well-known since [G]. However, we shall start with the introduction of the Liouville integral for the equation

$$(1.2) \qquad -\Delta u = \lambda e^u \quad \text{in} \quad \Omega \subset \mathbf{R}^2.$$

In fact, through a complex transformation of variables the equation (1.1) is transformed into the liouville equation

$$\frac{\partial^2}{\partial x \partial y} \log f = \pm f / 2a^2$$

of which integral with two arbitrary functions was discovered in 1853 ([L]). Since u is real-valued, the integrability is expressed in the following way ([B1]), ([NS2]):

PROPOSITION 1.1. The function $u = u(x) \in C^2(\Omega)$ solves (1.2) if and only if there exists an analytic function $f = f(z)$ in $\Omega \subset C$ in the sense of Weierstrass such that $\rho(f) =\equiv |f'|/(1+|f|^2)$ is positively single-valued and

$$(1.3) \qquad \left(\frac{\lambda}{8}\right)^{1/2} e^{u/2} = \rho(f),$$

where $z = x_1 + ix_2$ for $x = (x_1, x_2)$.

We should note that the function $f = f(z)$ is not uniquely determined by the solution u (c.f. [M]). Anyway Proposition 1.1 reduces to find an analytic function $f = f(z)$ so that

$$(1.4) \qquad \rho(f) = \left(\frac{\lambda}{8}\right)^{1/2} \quad (\text{on } \partial\Omega).$$

On the other hand, radial solutions for (1.1) on $D = \{|x| < 1\} \subset \mathbf{R}^2$ have been classified ([G]). In fact the equation for $u = u(r)$:

$$(1.5) \qquad (ru') + \lambda r e^u = 0 \quad \text{for} \quad r > 0 \quad \text{and} \quad u'(0) = 0$$

is integrable and we obtain.

PROPOSITION 1.2. ([G]). When $\Omega = D \subset \mathbf{R}^2$, the problem (1.1) has solutions only for $\lambda \in (0,2)$. For these λ's, the solutions are expressed explicitly by

$$(1.6) \qquad u_\pm(r) = \log \frac{\frac{8}{\lambda}C_\pm}{(1+C_\pm r^2)^2} \quad \text{with} \quad C_\pm = \frac{1}{\lambda}\{4 - \lambda \pm 2\sqrt{4-2\lambda}\},$$

where $r = |x|$.

Recall the diagram given in §1 of Part I. Thus, for this case we can take $f(z) = C^{1/2}z$ with $C = C_\pm$ in the associated Liouville integral (1.3).

Actually, the quantity $\rho(f) = \frac{|f'|}{1+|f|^2}$ has the following geometrical meaning ([NS]). Namely, let K be the Riemann sphere with unit diameter, tangent to w-plane at the origin, where $w = f(z)$. Further, let $d\sigma \doteq |dz|$ and $d\tau$ denote respectively the natural linear elements on Ω and on K induced by $\bar{f} : \Omega \to K$, the natural conformal mapping associated with f. Then they satisfy the relation $d\tau = \rho(f)\,d\sigma$. In this terminology, (1.4) yields that the induced length of $\bar{f}(\partial\Omega)$ on K as an immersion is equal to

$$(1.7) \qquad \Lambda \equiv \int_{\partial\Omega} \frac{|f'|}{1+|f|^2} d\sigma = \left(\frac{\lambda}{8}\right)^{1/2} |\partial\Omega|,$$

where $|\partial\Omega|$ denotes the lenght of $\partial\Omega$. Accordingly, the area of $\bar{f}(\Omega)$ on K as an immersion is given by

$$\Sigma \equiv \int_\Omega \left(\frac{|f'|}{1} + |f|^2\right)^2 \mathrm{d}x = \frac{\lambda}{8}\int_\Omega e^u \mathrm{d}x. \tag{1.8}$$

When $\Omega = D$, for a solution (λ, u) of (1.1) we define a number S by

$$S = 8\Sigma = \lambda \int_\Omega e^u \mathrm{d}x. \tag{1.9}$$

Then, all solutions (λ, u) of (1.1) can be parametrized by $S \in (0, 8\pi)$.

THEOREM 3. ([SN1, 2]). When $\Omega = D$, there is a continuous and one-to-one correspondence between $S \in (0, 8\pi)$ in (1.9) and the solution $(\lambda, u) \epsilon \mathbb{R}^+ \times C(\bar{\Omega})$ of (1.1). Denoting the solution corresponding to S by (λ_s, u_s), we have

$$\lambda_S = \frac{1}{8\pi^2} S(8\pi - S), \tag{1.10}$$

$$m(S) \equiv \int_\Omega e^{u_S} \mathrm{d}x = \frac{S}{\lambda_S} = \frac{8\pi^2}{8\pi - S} \tag{1.11}$$

and

$$n(S) \equiv \int_\Omega |\nabla u_S|^2 \mathrm{d}x = -16\pi \log(8\pi - S)\{1 + o(1)\} \quad S \nearrow 8\pi. \tag{1.12}$$

PROOF. The first part is obvious from the geometrical viewpoint. Actually, the area Σ of $\bar{f}(D)$ increases from 0 to π along the solution branch for (1.1) starting from $(\lambda, u) = (0,0)$. See [SN1, 2] for details and analytic proof. We shall show the relations (1.10) - (1.12). In fact, from (1.6) we have

$$S = 2\pi(2 \pm \sqrt{4 - 2\lambda})$$

and hence (1.10) and (1.11) follow. On the other hand, we have

$$\int_\Omega |\nabla u_\pm|^2 \mathrm{d}x = \lambda \int_\Omega u_\pm e^{u_\pm} \mathrm{d}x = 8\pi\left\{\log \frac{8C_\pm}{\lambda} - \frac{C_\pm}{C_\pm + 1}\right\}.$$

Now, $S/8\pi$ corresponds to $C = C_+ \sim \frac{8}{\lambda}$ and hence we have

$$n(S) = 8\pi \log\left(\frac{8}{\lambda}\right)^2 + 0(1) = -16\pi \log(8\lambda - S)(1 + o(1))$$

by (1.10).

2. Structure of radial solutions

It may look strange but only recently radial solutions for (1.1) in the annulus $\Omega = A \equiv \{a < |x| < 1\} \subset \mathbf{R}^2$ are classified, in spite that they are given explicitly ([SN2, 3], [Lin]). Namely, they are associated with the liouville integral (1.3) for $f(z) = \beta^{1/2} z^\alpha$, where the parameters $\alpha > 0$ and $\beta > 0$ are to be determined so that (1.4), i.e.,

$$\frac{\alpha\beta^{1/2}}{1+\beta} = \left(\frac{\lambda}{8}\right)^{1/2} \quad \text{and} \quad \frac{\alpha\beta^{1/2} a^{\alpha-1}}{1+\beta a^{2\alpha}} = \left(\frac{\lambda}{8}\right)^{1/2}. \tag{2.1}$$

Putting $\zeta = a^\alpha$, we can reduce the relation (2.1) to

$$\frac{\lambda}{8} a^2 (\log a)^2 = G(\zeta) \equiv \frac{(\log \zeta)^2 (a-\zeta)\zeta(1-a\zeta)}{(1-\zeta^2)^2} \tag{2.2}$$

with

$$\beta = \frac{a-\zeta}{\zeta(1-a\zeta)}. \tag{2.3}$$

We shall classify the radial solutions for (1.1) in A, studying the equation (2.2), and on the other hand showing that there are no other radial solutions.

The latter part is done by characterizing the solutions for

$$(rv')' + \lambda r e^v = 0 \quad \text{with} \quad v(a) = 0 \quad \text{and} \quad v'(a) > 0 \tag{2.4}$$

by the behaviour near $r = 0$. Actually, we have the following two propositions ([SN2]):

PROPOSITION 2.1. Each solution $v = v(r)$ of (2.4) continues up to $r = 0$ and for some real numbers $L > 0$ and M satisfies

$$\lim_{r \downarrow 0} \{v(r) - L \log r\} = M \quad \text{and} \quad \lim_{r \downarrow 0} \{v'(r) - \frac{L}{r}\} = 0. \tag{2.5}$$

PROPOSITION 2.2. For each numbers $L > 0$ and M, there exists a unique solution $v = v(r)$ of (2.4) satisfying the condition (2.5). If we take $f(z) = \beta^{1/2} z^\alpha$ for

$$\alpha = 1 + \frac{L}{2} \quad \text{and} \quad \beta = \lambda M / 2(2+L)^2, \tag{2.6}$$

then the function $v = v(r)$ defined through (1.3), i.e.,

$$v(r) = \log \frac{8}{\lambda}\alpha^2 \beta r^{2(\alpha-1)}/(1+\beta r^{2\alpha})^2 \tag{2.7}$$

satisfies (2.5) with $(rv')' + \lambda r e^v = 0$. Thus, through (2.1), radial solutions for (1.1) on $A = \{a < |x| < 1\}$ are to be classified.

Now, we shall describe some properties of (2.2). The function $G = G(\zeta)$ is positive in $(0, a)$ and satisfies $\lim_{\zeta \downarrow 0} G(\zeta) = G(a) = 0, \lim_{\zeta \downarrow 0} G'(\zeta) = +\infty$ and $G'(a) < 0$. Further,

PROPOSITION 2.3. (c. f. [SN2]). When $a \in (0,1), G(\zeta)$ has only one critical point in $(0, a)$.

Thus, we have the following theorem, where m^* denotes the maximum value of $G(\zeta)$ in $(0, a)$ and put $\lambda^* = 8m^*/a^2(\log a)^2$.

THEOREM 4. ([NS2], [L]). There exists a positive number λ^* such that the number of radial solutions for (1.1) on $\Omega = A \equiv \{a < |x| < 1\} \subset \mathbf{R}^2$ is two, one and zero for $\lambda \in (0, \lambda^*), \lambda = \lambda^*$ and $\lambda \in (\lambda^*, \infty)$, respectively.

Every radial solution of (1.1) corresponds uniquely and continuosly to $\zeta \in (0, a)$, and hence to $\alpha \in (1, \infty)$. For $\lambda \in (0, \lambda^*)$, there are just two roots $\overline{\zeta}$ and $\underline{\zeta}$ of (2.2): one near 0 and the other near a. The corresponding solutions $\overline{v}_\lambda$ and $\underline{v}_\lambda$ of (1.1) will turn out to be the «large» and minimial solutions, respectively.

PROPOSITION 2.4. (SN [2]). We have

$$\begin{aligned} &\lim_{\lambda \downarrow 0} \underline{v}_\lambda(x) = 0 \quad \text{and} \\ &\lim_{\lambda \downarrow 0} \overline{v}_\lambda(x) = +\infty \quad \text{for each} \quad x \in A \end{aligned} \tag{2.8}$$

and

$$\begin{aligned} &p_-(r) \equiv \lambda e^{\underline{v}} \to 0 \quad \text{and} \\ &p_+(r) \equiv \lambda e^{\overline{v}} \to \begin{cases} 0 (r \neq \sqrt{a}) \\ \\ +\infty (r = \sqrt{a}) \end{cases} \quad \text{as } \lambda \downarrow 0. \end{aligned} \tag{2.9}$$

Thus, the large radial solution $\overline{v}_\lambda$ makes the entire blow-up as $\lambda \downarrow 0$. We recall that for the disc $\Omega = D$, the large solution makes one-point blow up as $\lambda \downarrow 0$, i.e., $u_+(r) \sim 4 \log \frac{1}{r}$ as $\lambda \downarrow 0$.

Further in Theorem 3, all the solutions of (1.1) on $\Omega = D$ are parametrized by $S = 8\Sigma$, where Σ in (1.8) represents the area of $\bar{f}(\Omega)$ on the Riemann sphere K for $f(z) = C^{1/2}z$. On the other hand, for the case $\Omega = A$ we have taken $f(z) = \beta^{1/2}z^{\alpha}$ which is α-fold on $\Omega = A$. Accordingly, the real area of $\bar{f}(\Omega)$ on K is given by $\frac{1}{\alpha}\Sigma$ this time. These facts suggest us that every radial solution (λ, u) of (1.1) on the annulus A may be parametrized by $\frac{\lambda}{\alpha}\int_{\Omega} e^{u} dx$. This fact has been proven, in fact, for the solutions $(\lambda, \bar{v}_{\lambda})$ with $\lambda > 0$ small enough.

To state the result, we introduce the parameter σ for large radial solutions $(\lambda, \bar{v}_{\lambda})$ of (1.1) by

$$\sigma \equiv \frac{\lambda}{\alpha}\int_{\Omega} e^{\bar{v}_{\lambda}} dx. \tag{2.10}$$

PROPOSITION 2.5. The mapping $\zeta \to \sigma$ is one-to-one when ζ is near to 0. Moreover,

$$\sigma = 8\pi\left\{1 - \left(a + \frac{1}{a}\right)\zeta + o(\zeta^2)\right\} \quad \text{as} \quad \zeta \downarrow 0 \tag{2.11}$$

and hence

$$\zeta = \frac{1}{8\pi}\left(a + \frac{1}{a}\right)^{-1}(8\pi - \sigma)\{1 + o(8\pi - \sigma)\} \quad \text{as} \quad \sigma \nearrow 8\pi. \tag{2.12}$$

PROOF. Putting $\kappa = \left(\frac{8}{\lambda}\right)^{1/2}\alpha\beta^{1/2} = 1 + \beta$, we have

$$\bar{v}_{\lambda}(r) = -2\log\left\{A(\zeta)\left(\frac{a}{r}\right)^{\alpha-1} + B(\zeta)r^{\alpha+1}\right\} \tag{2.13}$$

with

$$A(\zeta) = \frac{1}{\kappa a^{\alpha-1}} = \frac{1 - a\zeta}{1 - \zeta^2} = 1 - a\zeta + 0(\zeta^2) \tag{2.14}$$

and

$$B(\zeta) = \frac{B}{\kappa} = \frac{a - \zeta}{a(1 - \zeta^2)} = 1 - \frac{\zeta}{a} + 0(\zeta^2) \tag{2.15}$$

Therefore,

$$\sigma = \frac{\lambda}{\alpha}\int_{\Omega}\left\{A(\zeta)\left(\frac{a}{r}\right)^{\alpha-1} + B(\zeta)r^{\alpha+1}\right\}^{-2}\mathrm{d}x$$
$$= 16\pi\alpha C(\zeta)\int_{a}^{1}\left\{A(\zeta)\left(\sqrt{a}/r\right)^{\alpha-1} + aB(\zeta)(r/\sqrt{a})^{\alpha+1}\right\}^{-2} r\,\mathrm{d}r,$$

where $C(\zeta) = \frac{\lambda a}{8\alpha^2\zeta} = \frac{(a-\zeta)(1-a\zeta)}{a(1-\zeta^2)^2} = 1 - \left(a+\frac{1}{a}\right)\zeta + 1(\zeta^2)$. Substitution of $t = (r/\sqrt{a})^{\alpha}$ in the above integral yields

$$\sigma = 16\pi\alpha C(\zeta)\frac{a}{\alpha}\int_{\zeta^{1/2}}^{\zeta^{-1/2}}\{A(\zeta)t^{-1} + aB(\zeta)t\}^{-2}\frac{\mathrm{d}t}{t}$$
$$= 16\pi a C(\zeta)\left[-\frac{1}{2aB(\zeta)}(A(\zeta) + aB(\zeta)t^2)^{-1}\right]_{t=\zeta^{1/2}}^{t=\zeta^{-1/2}}$$
$$= \frac{8\pi a(1-\zeta^2)C(\zeta)}{\{A(\zeta) + aB(\zeta)\zeta\}\{A(\zeta)\zeta + aB(\zeta)\}}.$$

Here, we recall (2.14) and (2.15) to conclude (2.11).

The term $0(\zeta^2)$ in (2.11) is smooth and its derivative converges to 0 as $\zeta \downarrow 0$. Consequently, we have

$$\lim_{\zeta\downarrow 0}\frac{\mathrm{d}\sigma}{\mathrm{d}\zeta} = -8\pi\left(a+\frac{1}{a}\right),$$

from which the relation (2.12) is derived by the implicit function theorem.

Eventually we can regard $\mu \equiv \int_{\Omega} e^{\overline{v}_{\lambda}}\mathrm{d}x$ and $\nu \equiv \int_{\Omega} |\nabla\overline{v}_{\lambda}|^2\mathrm{d}x$ as functions of σ, namely $\mu = \mu(\sigma)$ and $\nu = \nu(\sigma)$. Their behavior for σ near to 8π is as follows:

PROPOSITION 2.6. We have

$$\mu(\sigma) = \frac{8\pi^2(a^2+1)\log a}{(8\pi-\sigma)\log(8\pi-\sigma)}\{1+o(1)\} \tag{2.16}$$

and

$$\nu(\sigma) = \frac{-8\pi}{\log a}\left(\log\frac{1}{8\pi-\sigma}\right)^2\{1+o(1)\} \quad \text{as} \quad \sigma \nearrow 8\pi. \tag{2.17}$$

PROOF. From (2.2) we have

$$\frac{\alpha}{\lambda} = \frac{a^2(\log a)(1-\zeta^2)}{8(\log\zeta)(a-\zeta)\zeta(1-a\zeta)} = \frac{a\log a}{8\zeta\log\zeta}\{1+\left(a+\frac{1}{a}\right)\zeta + 0(\zeta^2)\},$$

and hence

$$\mu(\sigma) = \frac{\alpha}{\lambda}\cdot\sigma = \frac{\pi a\log a}{\zeta\log\zeta}\{1+0(\zeta^2)\}$$

by (2.11) as $\zeta\downarrow 0$. Hence (2.16) follows.

On the other hand we have

$$\begin{aligned}
\nu(\sigma) &= -\int_\Omega \overline{v}_\lambda \Delta\overline{v}_\lambda \mathrm{d}x = \lambda\int_\Omega e^{\overline{v}_\lambda}\overline{v}_\lambda \mathrm{d}x = \\
&= -4\pi\lambda\int_a^1\left\{A(\zeta)\left(\frac{a}{r}\right)^{\alpha-1}+B(\zeta)r^{\alpha+1}\right\}^{-2}\log\left[A(\zeta)\left(\frac{a}{r}\right)^{\alpha-1}+\right. \\
&\left.+B(\zeta)r^{\alpha+1}\right]\cdot r\mathrm{d}r = \\
&= -\frac{4\pi\lambda a}{\zeta}\int_a^1\{A(\zeta)(\sqrt{a}/r)^{\alpha-1}+aB(\zeta)(r/\sqrt{a})^{\alpha-1}\}^{-2}\times \\
&\times\log\left[A(\zeta)\left(\frac{a}{r}\right)^{\alpha-1}+B(\zeta)r^{\alpha+1}\right]\times \\
&\times\{\log a^{\frac{\alpha}{2}-1}r+\log[A(\zeta)(\sqrt{a}/r)^\alpha+aB(\zeta)(r/\sqrt{a})^\alpha]\}r\mathrm{d}r.
\end{aligned}$$

Here, substitution of $t=(r/\sqrt{a})^\alpha$ as before yields

$$\begin{aligned}
\nu(\sigma) &= -\frac{4\pi\lambda\alpha^2}{a\zeta}\int_{\zeta^{1/2}}^{\zeta^{-1/2}}\{A(\zeta)+aB(\zeta)t^2\}^{-2}\times \\
&\times\left\{\frac{\alpha-1}{2}\log a+\frac{1-\alpha}{\alpha}\log t+\log[A(\zeta)+aB(\zeta)t^2]\right\}t\mathrm{d}t\equiv \\
&\equiv -\frac{4\pi\lambda a^2}{\alpha\zeta}\{I+II+III\}.
\end{aligned}$$

We have

$$\begin{aligned}
I &= \frac{\alpha-1}{2}\log a\left[-\frac{1}{2aB(\zeta)}(A(\zeta)+aB(\zeta)t^2)^{-1}\right]_{t=\zeta^{+1/2}}^{t=\zeta^{-1/2}} \\
&= \frac{\alpha-1}{2a}(\log a)\{1+0(\zeta^2)\}.
\end{aligned}$$

Substituting $\tau = B(\zeta)A(\zeta)^{-1}at^2$, we have

$$II = \frac{1-\alpha}{2\alpha}A(\zeta)(B(\zeta)a)^{-1}\int_{\tau_-}^{\tau_+}(1+\zeta)^{-2}\{\log\tau-\log B(\zeta)A(\zeta)^{-1}a\}\mathrm{d}\tau$$

where $\tau_{\pm} = B(\zeta)A(\zeta)^{-1}a\zeta^{\pm 1}$.

Nothing that $A(\zeta), B(\zeta) \to 1$ and $\alpha \to +\infty$ as $\zeta \downarrow 0$, we have

$$II = -\frac{1}{2a}\int_o^\infty (1+\tau)^{-2}\{\log\tau - \log a\}\mathrm{d}\tau + o(1) = 0(1).$$

Finally, throught the same substitution we have

$$III = \frac{1}{a}\int_0^\infty \frac{\log(1+\tau)}{(1+\tau)^2}\mathrm{d}\tau + o(1) = 0(1).$$

Thus, we get

$$\nu(\sigma) = -\frac{\pi\lambda a\log a}{a}\{1+o(1)\} = -\frac{8\pi(\log\zeta)^2}{\log a}\{1+o(1)\}$$

as $\zeta \downarrow 0$.

We can summarize the preceding results as

THEOREM 5. ([SN2]). The large radial solutions $(\lambda, \overline{v}_\lambda)$ for (1.1) in $\Omega = A$ are parametrized by σ close to 8π with the properties (2.16) and (2.17) when $\lambda > 0$ is small.

3. Generation of non-radial solutions.

From variational approach, we shall show the existence of non-radial solutions for the nonlinear eigenvalue problem (1.1) in $\Omega = A$. We first describe our formulation of variation for the general semilinear eigenvalue problem

$$(3.1) \qquad -\Delta u = \lambda f(u) \quad (\text{in } \Omega), \quad u = 0 \quad (\text{on } \partial\Omega),$$

where $\lambda \in \mathbb{R}$ and $\Omega \subset \mathbb{R}^n$ is a smooth bounded domain.

For this purpose we assume that a continuous function $f : R \to \mathbb{R}$ satisfies for every $t \in \mathbb{R}$ that

$$(3.2) \qquad |f(t)| \leq \begin{cases} C(1+|t|^p) & \text{for } 1 < p < (n+2)/(n-2) \text{ when } n \geq 3 \\ C\exp|t|^q & \text{for } 0 < q < 2 \text{ when } n = 2; \end{cases}$$

where $C > 0$. Let K be a closed subspace of $v = H_0^1(\Omega)$ having the property that for every $v \in K$ the solution $w \in H_0^1(\Omega)$ of the problem

$$(3.3) \qquad \Delta w + f(v) = 0 \quad \text{in} \quad \Omega,\ w = 0 \quad \text{on} \quad \partial\Omega$$

also belongs to K. Later we shall call this the invariance of K respect to (3.3). Introducing two functionals $\phi(v) = \int_\Omega F(v)\,dx$ and $J(v) = \frac{1}{2}\int_\Omega |\nabla v|^2\,dx$ for $v \in H_0^1(\Omega)$, where $F(t) = \int_0^t f(S)\,dS$, we define subsets K_c of K by

$$K_c = \{v \in K \mid \phi(v) = c\} \quad \text{for} \quad c \in \mathbf{R}$$

and non-negative numbers j_c by

$$j_c = \text{Inf}\,\{J(v) \mid v \in K_c\} \quad \text{when} \quad K_c \neq \emptyset.$$

Under these circumstances we have

PROPOSITION 3.1. (c.f. [NS3]). There exists a minimizer $v_c \in K_c$ of j_c. If there does not exist a $v \in V$ such that $\phi(v) = c$ with $f(v) = 0$, we have with some Langrangen multiplier λ_c that

$$\text{(3.4)} \qquad -\Delta v_c = \lambda_c f(v_c) \quad (\text{in } \Omega) \quad v_c = 0 \quad (\text{on } \partial\Omega).$$

PROOF. The existence of a minimizer v_c of j_c is quite known under the assumption (3.2) (C.f. [R]). the functionals ϕ and ζ have Fréchet derivatives $\phi'(v)$ and $J'(v)$ given by

$$\phi'(v)\varphi = \int_\Omega f(v)\varphi\,dx \quad \text{and} \quad J'(v)\varphi = \int_\Omega \nabla v \cdot \nabla\varphi\,dx$$

for $v, \varphi \in V = H_0^1(\Omega)$. Therefore, from the general theory of extremal problems, there exist Langrangen multipliers μ and ν with $(\mu, \nu) \neq (0, 0)$ such that

$$\text{(3.5)} \qquad \mu\int_\Omega \nabla v_c \cdot \nabla\varphi\,dx + \nu\int_\Omega f(v_c)\varphi \in K.$$

We first show that $\mu \neq 0$. In fact, assume $\mu = 0$ and take the solution $w \in V = H_0^1(\Omega)$ in (3.3) for $v = v_c$. Since $v_c \in K$ we have $w \in K$ from the invariance. Therefore, we can take $\varphi = w$ in (3.5). Now $\mu = 0$ and hence $\nu \neq 0$ so that

$$0 = \int_\Omega (-\Delta w) \cdot w\,d\,wx = \int_\Omega |\nabla w|^2\,dx.$$

Hence $w = 0$ and consequently $f(v_c) = 0$, which contradicts to $\{v \in V \mid f(v) = c, \phi(v) = 0\} = \emptyset$.

Therefore in the terminology of subdifferentials ([Br]), (3.5) reads as

$$\lambda_c f(v_c) \in \partial(\phi + 1_k)(v_c) \quad \text{with} \quad v_c \in K, \tag{3.6}$$

where $\lambda_c = -\nu/\mu$, which means for $f = \lambda_c f(v_c) \in V^*$ and $u = v_c \in K$ that

$$\phi(\zeta) \geq \phi(u) +_{V^*} \langle f, \zeta - u\rangle_V \tag{3.7}$$

for each $\zeta \in K$. Furthermore, from the invariance we have a $W \in K$ such that

$$\phi(\zeta) \geq \phi(w) +_{V^*} \langle f, \zeta - w\rangle_V \tag{3.8}$$

for each $\zeta \in V$. Taking $\zeta = w$ in (3.7), we add it to (3.8) to obtain

$$\phi(\zeta) \geq \phi(u) +_{V^*} \langle f, \zeta - w\rangle_V$$

for each $\zeta \in V$ and hence $f \in \partial\phi(u)$. This means (3.4).

Now we apply the proposition for $f(t) = e^t, \Omega = A = \{a < |x| < 1\} \subset \mathbf{R}^2$ and $K = V_k$ defined in Part I. Then the invariance principle holds. Introduce the functionals $J(v) = \frac{1}{2}\int_\Omega |\nabla v|^2 \mathrm{d}\,s$ and $M(v) = \int_\Omega e^v \mathrm{d}\,x$, and consider the variational problem

$$j_k(\mu) = \mathrm{Inf}\,\{J(v)|v \in V_{k,\mu}\}, \tag{3.9}$$

where $V_{k,\mu} = \{v \in V_k | M(v) = \mu\}$. Then there exists a minimizer $u = u_{k,\mu} \in V_{k,\mu}$ satisfying

$$-\Delta u = \lambda e^u \quad (\text{in } A), \quad u = 0 \quad (\text{on } \partial A) \tag{3.10}$$

for some $\lambda \in \mathbf{R}$. When $\mu > |\Omega|$, the relation $\lambda > 0$ holds. In fact, if $\lambda \leq 0$ then we have $u \leq 0$ so that $\mu = \int_\Omega e^u \mathrm{d}\,x \leq |\Omega|$.

Here we claim the following proposition from which generation of non-radial solutions for (1.1) is assured.

PROPOSITION 3.2. For any $k = 1, 2, \ldots,$ the inequality

$$j_\infty(\mu) > j_k(\mu) \tag{3.11}$$

holds if μ is loarge enough.

PROOF. As is seen in §2, the solution $(\lambda, V_\infty) \in K_{\infty,\mu}$ is unique for μ large enough. Here the infinimum $j_\infty(\mu)$ of $J(v)$ on $K_{\infty,\mu}$ is attained by $v = v_\infty$. Moreover, (λ, V_∞) is parameterized by some parameter $\sigma \in (0, 8\pi)$ with the relations

$$\mu = \mu(\sigma) = \frac{8\pi^2(a^2+1)\log a}{(8\pi-\sigma)\log(8\pi-\sigma)}\{1+o(1)\} \tag{3.12}$$

and

$$j_\infty(\mu) = \frac{1}{2}\nu(\sigma) = \frac{-4\pi}{\log a}\left(\log\frac{1}{8\pi-\sigma}\right)^2\{1+o(1)\} \tag{3.13}$$

as $\sigma \nearrow 8\pi$. Therefore, μ large enough determines σ close to 8π via (3.12).

Now we turn to estimate $j_k(\mu)$ for $k < \infty$. To this end we shall make use of the radial solutions for $\Omega \equiv D$.

Take a disk ω_k in the fundamental region $\Omega_k = \left\{re^{i\theta} | a < r < 1, 0 < \theta < \frac{2\pi}{k}\right\}$ with a center x_0 and radius ϵ and define the function $U \in H_0^1(\Omega_k)$ by

$$U(x) = u_s\left(\frac{1}{\epsilon}(x - x_0)\right) \quad \text{in} \quad \text{and} \quad U(x) = 0 \quad \text{in} \quad \Omega\backslash\omega_k,$$

where $u_s = u_s(x)$ $(S \in (0, 8\pi))$ is the parametrized solution for $\Omega = D$ in Theorem 3. We put

$$w_s(x) = U(x) + U(T_k x) + U(T_k^2 x) + \ldots + U(T_k^{k-1} x)$$

for $x \in \Omega$. Then, $w_S \in V_k$ follows. Through a simple calculation, we get for this w_s that

$$M(w_s) = k\epsilon^2 m(s) + |\Omega| - k\epsilon^2\pi \tag{3.14}$$

and

$$J(w_s) = \frac{1}{2}kn(s), \tag{3.15}$$

with $m(s)$ and $n(s)$ in (1.11) and (1.12), respectively. Therefore, similarly to the case of σ, μ large enough determines s near to 8π uniquely by $M(w_c) = \mu$, in which case $w_s \in V_{k,\mu}$.

Accordingly, σ and s correspond the relation

$$k\epsilon^2\frac{8\pi^2}{8\pi - s} + |s| - k\epsilon^2\pi = \frac{8\pi^2(a^2+1)\log a}{(8\pi-\sigma)\log(8\pi-\sigma)}\quad\{1+o(1)\}. \tag{3.16}$$

Then we have

$$j_k(\mu) \leq J(w_s) = 8\pi k \left(\log \frac{1}{8\pi - \sigma}\right) \{1 + o(1)\} \tag{3.17}$$

as $\sigma \nearrow 8\pi$. Hence (3.11) holds by (2.17) and (3.17).

Once Proposition 3.2 has been established, through the argument Proposition 2.3 in Part I we can separate $\{j_k(\mu)\}$ for finite k's. Thus we obtain

THEOREM 6. ([SN2]). For any positive integer k, there exists a number μ_k such that for any $\mu > \mu_k$ the problem (1.1) has a non-radial solution (λ, u) of mode k satisfying $M(u) = \mu$.

4. Remark on symmetry-breaking

It would be interesting how each family of k-mode solutions in Theorem 6 exists in $\lambda - u$ plane. We expected that if bifurcates from the branch of radial solutions and reduced the problem to asymptotic analysis for some associated Legendre equations ([SN3]). We could not see the fact then, but the solutions for those equations have explicit forms. Recently, [Lin] has independently studied the problem and proved the fact for their equivalent equations. Hence now we can say that symmetry-breaking actually occurs. Those bifurcating cones, i.e., two- dimensional manifolds in $\lambda - u$ plane, are shown to be only infinitesimally k-mode up to now. However, we except that they are globally k-mode and coincide with the families of solutions in Theorem 6. In this section, we shall review the result [Lin] briefly utilizing the associated Legendre equations.

Let $\{(\lambda, \overline{v}_\lambda)\}$ be the branch of large symmetric solutions in $A = \{a < |x| < 1\} \subset \mathbb{R}^2$ for (1.1). We shall examine the degeneracy of the linearized operator $-\Delta - p(r)$ under Dirichlet condition, where

$$p(r) = \lambda e^{\overline{v}_\lambda(r)} = \frac{8\alpha^2 a^{-1} j}{r^2 \left[\left(\frac{r}{\sqrt{a}}\right)^\alpha + a^{-1} j \left(\frac{\sqrt{a}}{r}\right)^\alpha\right]^2} \quad \text{with} \quad j = \frac{1 - a\zeta}{1 - a^{-1}\zeta}.$$

Noting the 0 (2)-symmetry of (1.1), the problem is then reduced to the bifurcation from simple eigenvalues, where the theory of [R] or [CR] works. See [SW], for instance.

In view of the separation of variables, the linearized equation

$$-\Delta\varphi = \frac{p(r)}{\Lambda}\varphi \quad (\text{in } A), \quad \varphi = 0 \quad (\text{on } \partial A) \tag{4.1}$$

is nothing but

$$(4.2)\quad \frac{1}{r}(r\psi_r)_r + [p(r)/\Lambda - k^2/r^2]\psi = 0 \quad (a < r < 1) \quad \text{with} \quad \psi|_{r=a,1} = 0.$$

where $k = 0, 1, 2, \ldots$ Introducing the variable $t = \left(\frac{r}{\sqrt{a}}\right)^{\alpha}$, we get

$$(4.3)\quad \begin{gathered} \frac{1}{t}(t\psi_t)_t + [q(t)/\Lambda - \left(\frac{k}{\alpha}\right)^2 /r^2]\psi = 0 \\ \left(\zeta^{-1/2} < t < \zeta^{1/2}\right) \quad \text{with} \quad \psi|_{t=\zeta^{\pm 1/2}} = 0, \end{gathered}$$

where $q(t) = \dfrac{8a^{-1}j}{(t^2 + a^{-1}j)^2}$. Now we introduce the Bandle transformation $\xi = (a^{-1}j - t^2)/(a^{-1}j + t^2)$, putting $\mu = \frac{k}{\alpha}$, to obtain

$$(4.4)\quad \begin{gathered} [(1-\xi^2)\psi_\xi]_\xi + \left[\frac{2}{\Lambda} - \mu^2/(1-\xi^2)\right]\psi = 0 \\ (\xi_- < \xi < \xi_+) \quad \text{with} \quad \psi|_{\xi=\xi_\pm} = 0, \end{gathered}$$

where $\xi_+ = \dfrac{a^{-1}j - \zeta}{a^{-1}j + \zeta}$ and $\xi_- = \dfrac{a^{-1}j - \zeta^{-1}}{a^{-1}j + \zeta^{-1}}$.

Now we claim that

PROPOSITION 4.1. The equation

$$(4.5)\quad [(1-\xi^2)\psi_\xi]_\xi + [2 - \mu^2/(1-\xi^2)]\psi = 0 \quad (-1 < \xi < 1)$$

has the following fundamental solutions:

Case 1.

$$\mu \neq 0, \pm 1; \varphi_1 = (1-\xi)^{\frac{\mu}{2}}(1+\xi)^{-\frac{\mu}{2}}(\xi + \mu),$$
$$\varphi_2 = (1-\xi)^{-\frac{\mu}{2}}(1+\xi)^{\frac{\mu}{2}}(\xi - \mu).$$

Case 2.

$$\mu = 0;\ \varphi_1 = \xi,\ \varphi_2 = -1 + \frac{\xi}{2}\log\frac{1+\xi}{1-\xi}$$

Case 3.

$$\mu = \pm 1 : \varphi_1 = (1-\xi^2)^{1/2},\ \varphi_2 = \left\{\log\frac{1+\xi}{1-\xi} + \frac{2\xi}{1-\xi^2}\right\}(1-\xi^2)^{1/2}.$$

Hence the degeneracy of linearized operators is reduced to the algebraic equation

$$\frac{\varphi_2}{\varphi_1}(\xi_+) = \frac{\varphi_2}{\varphi_1}(\xi_-). \tag{4.6}$$

See [Lin], for the analysis of this equation.

We note that [B1] has known the solution $\varphi_1 = \xi$ for $\mu = 0$. Then, the other solution $\varphi_2 = -1 + \frac{\xi}{2} \log \frac{1+\xi}{1-\xi}$ can be obtained by Wronskian's method. Calculating $(1-\xi^2)^{\frac{n}{2}} \left(\frac{d}{d\xi}\right)^n \varphi_2$, we get $\varphi_1 = (1-\xi)^{-\frac{\mu}{2}}(1+\xi)^{\frac{\mu}{2}}(\xi - \mu)$ with $\mu - n (= 1, 2, \ldots)$. Hence the fundamental system of solutions for $\mu = n (\geq 2)$ arise, which happen to be the cases for $\mu \neq 0, \pm 1$, too. [Lin] has discovered those solutions independently through the equivalent equation

$$(1+s)^2 (s\varphi_s)_s + \{2 - \mu^2 (1+s)^2 / 4s\} \varphi = 0. \tag{4.7}$$

REFERENCES

[A] A. D. ALEXANDROV and H. HOPF: *Differential Geometry in the Large.* Lecture Notes in Math., 1000, Springer, Berlin Heidelber New York Tokyo, 1983.

[B1] C. BANDLE: *Isoperimetric Inequalities and Applications.* Pitman, Boston London Melbourne, 1980.

[B2] C. BANDLE: Solutions for a nonlinear Dirichlet problem for nearly circular domains. *SIAM J. Numer. Anal.*, **20**, 1983, 1094-1098.

[Br] H. BRÉZIS: *Opérateurs Maximaux Monotones et Semi-groupes de Contradictions dan les Espace de Hilbert.* North-Holland, Amsterdam London New York, 1973.

[Bn] H. BRÉZIS and L. NIRENBERG: Positive solutions of nonlinear elliptic equations involving critical Sobolev exponents. *Comm. Pure Appl. Math.*, **36**, 1983, 437-477.

[CNS] Y. G. CHEN, S. NAKANE and T. SUZUKI: Elliptic equations on 2D symmetric domains; local profile of mild solutions. Preprint, 1988.

[C] C.V. COFFMAN: A non-linear boundary value problem with many positive solutions. *J. Diff. Eqs.*, **54**, 1984, 429-437.

[Cr] M. G. CRANDALL and P. H. RABINOWITZ: Bifurcation from simple eigenvalues. *J. Func. Anal.*, **8**, 1971, 321-340.

[G] I. M. GELFAND: Some problems in the theory of quasilinear equations. *Amer. Math. Soc. Transl.*, **1 - 29**, 1963, 295-381.

[GNN] B. GIDAS, W-M. NI and L. NIRENBERG: Symmetry and related properties via the maximum principle. *Comm. Math. Phys.*, **68**, 1979, 209-243.

[JL] D. D. JOSEPH and T. S. LUNDGREN: Quasilinear Dirichlet problems driven by positive sources. *Arch. Rat. Mech. Anal.*, **49**, 1973, 241-269.

[K] B. KAWOHL: *Rearrangements and convexity of level sets in PDE.* Lecture Notes in Math., 1150, Springer, Berlin Heidelberg New York Tokyo, 1985.

[KW] J. L. KAZDAN and F. W. WARNER: Remarks on some quasilinear elliptic equations. *Comm. Pure Appl. Math.*, **28**, 1975, 567-597.

[Kc] H. B. KELLER and D. S. COHEN: Some positive problems suggested by nonlinear heat generation. *J. Math. Mech.*, **16**, 1967, 1361-1376.

[Lin] S. S. LIN: On non-radially symmetric bifurcation in the annulus. To appear in *J. Diff. Eqs.*

[L] J. LIOUVILLE: Sur l'equation aux différentes partialles $\partial^2 \log \lambda / \partial u \partial v \pm 2\lambda a^2 = 0$. *J. de Math.* **18**, 1853, 71-72.

[MMP] F. MIGNOT, F. MURAT and J. P. PUEL: Variation d'un point de retourment par rapport au domaine. *Comm. PDE*, **4**, 1979, 1263-1297.

[M] J. L. MOSELEY: Asymptotic solutions for a Dirichlet problem with an exponential nonlinearity. *SIAM J. Math. Anal.*, **14**, 1983, 719-735.

[Ns] K. NAGASAKI and T. SUZUKI: On a nonlinear eigenvalue problem. In: *Recent Topics in Nonlinear PDE III*, edited by K. Masuda and T. Suzuki, Kinokuniya/ North- Holland, Tokyo Amsterdam, 1987, 185-218.

[N] Z. NEHARI: On a nonlinear differential equation arising in nuclear physics. *Proc. Roy. Irish Acad.*, **62**, 1963, 117-135.

[R] P. H. RABINOWITZ: *Minimax Methods in Critical Point Theory with Applications to Differential Equations.* AMS, Providence, 1986.

[S] T. SUZUKI: Decreasing streamlines of solutions and spectral properties of linearized operators for semilinear elliptic equations. Preprint, 1988.

[SN1] T. SUZUKI and K. NAGASAKI: On the nonlinear eigenvalue problem $\Delta u + \lambda e^u = 0$. To appear in *Trans. AMS.*

[SN2] T. SUZUKI and K. NAGASAKI: Radial and non-radial solutions for the nonlinear eigenvalue problem $\Delta u + \lambda e^u = 0$ on annului, to appear in J. Diff. Eqs.

[SN3] T. SUZUKI and K. NAGASAKI: Lifting of local subdifferentiations and elliptic boundary value problems on symmetric domains, I and II. *Proc. Japan Acad.*, Ser. A., **64**, 1988, 1-4 and 29-34.

[SN4] T. SUZUKI and K. NAGASAKI: Nonlinear eigenvalue problem $\Delta u + \lambda e^u = 0$ on simply connected domains in $\mathbf{R}^2$. Proc. Japan Acad., Ser. A., 65, 1989, 74-76.

[SW] J. SMOLLER and A. WASSERMAN: Symmetry-breaking for positive solutions of semilinear elliptic equations. *Arch. Rat. Mech. Anal.*, **95**, 1986, 217-225.

FINITO DI STAMPARE IL 28 FEBBRAIO 1990
PRESSO LE OFFICINE GRAFICHE TECNOPRINT
VIA DEL LEGATORE 3, BOLOGNA (ITALY)